DISCOVER
TEXAS
DINOSAURS

Gulf Publishing Company
Houston, Texas

Discover Texas Dinosaurs

Charles E. Finsley

WITH SPECIAL CONSULTATION BY

DR. WANN LANGSTON, JR.

ILLUSTRATIONS BY DORIS TISCHLER

Discover Texas Dinosaurs

Gulf Publishing Company
Book Division
P.O. Box 2608 □ Houston, Texas 77252-2608

10 9 8 7 6 5 4 3 2 1

Library of Congress Cataloging-in-Publication Data
Finsley, Charles, 1938–
Discover Texas dinosaurs / Charles E. Finsley; with special consulting by Wann Langston, Jr.; illustrations by Doris Tischler.
p. cm.
Includes bibliographical references and index.
ISBN 0-87719-320-7 (alk. paper)
1. Dinosaurs—Texas. I. Title.
QE862.D5F52 1999
567.9′09764—dc21 98-47577
CIP

Printed on acid-free paper (∞).

Printed in Hong Kong.

Book design by Roxann L. Combs.

Cover design by Senta Eva Rivera.

Back cover photo of Charles E. Finsley with track of a Glen Rose meat-eating dinosaur, probably *Acrocanthosaurus,* by A. Barker.

Front cover photo of *Tenontosaurus* skull on exhibit, Dallas Museum of Natural History by C. Finsley.

Contents

Tenontosaurus currently on exhibit at the Dallas Museum of Natural History.

Acknowledgments

I would not have attempted this project without the considerable help of my major consultant, Dr. Wann Langston, Jr., University of Texas at Austin. He did his utmost to make the book paleontologically accurate. Please charge any mistakes strictly to my account. Working with me, Dr. Langston found his job was as much to teach as to advise.

Few artists have Doris Tischler's sensitivity to the biology behind putting the flesh back on dinosaur bones. She has worked with the finest paleontologists in the world, and has learned to interpret their suggestions. Her work graces many museum exhibits and publications. She illustrated my previous book, *A Field Guide to Fossils of Texas,* also by Gulf Publishing Co.

The major museums of Texas made me welcome to photograph their dinosaur exhibits: The Houston Museum of Natural Science, the Witte Museum in San Antonio, the Texas Memorial Museum in Austin, the Strecker Museum at Baylor University in Waco, the Dallas Museum of Natural History, the Fort Worth Museum of Science and History, the Shuler Museum of Paleontology at Southern Methodist University in Dallas, the Museum at Texas Tech University in Lubbock, the Panhandle-Plains Historical Museum in Canyon, Texas, the Oklahoma State Museum of Natural History in Norman, Oklahoma, and Big Bend National Park.

Curator of Paleontology Dr. Richard Cifelli and Exhibits Technician Kyle Davies (Oklahoma University); Walter R. Davis (Panhandle-Plains); Drs. Sankar Chatterjee and Thomas Lehman (Texas Tech); Drs. Louis and Bonnie Jacobs, Dr. Dale Winkler (Southern Methodist University); Mr. Jim Diffily (Fort Worth Museum of Science and History); Dr. Anthony Fiorillo and Fossil Preparator Geb Bennett (Dallas Museum of Natural History); Mr. David Lintz and Director Calvin Smith (Strecker Museum); Dr. Ernest L. Lundelius, Jr., Dr. Timothy Rowe, Dr. Wann Langston, Jr., Collection Manager Dr. Melissa Winans, and highly skilled Fossil Preparators Earl Yarmer and Robert Rainey (Texas Memorial Museum); Exhibits Curator, Hank Harrison and V. P. of Marketing

Lynda Alston (Witte Museum), and Delia Murta and Ivan Perez (Houston Museum of Natural Science)—all helped me to photograph and glean dinosaur insights at their institutions.

A special thanks to Gaylon Polatti, of the Dallas Historical Society Library and Archives, for Texas Centennial Exposition photos. Similar thanks to the Smithsonian Institution Archives, Washington, D.C., and the American Museum of Natural History, New York, for permission to use their photographic materials.

In addition, Dr. Dale Russell at North Carolina State University; Dr. Jeffrey Pittman at the University of Colorado at Denver; Dr. James Farlow at Indiana University-Purdue University at Fort Wayne; Dr. Catherine Forster at the State University of New York at Stony Brook; the late Bob Slaughter of SMU's Shuler Museum; Dr. Chris Scotese at the University of Texas at Arlington; Dr. John Thurman (retired) from the University of Arkansas at Little Rock; Dr. David Rohr at Sul Ross State University, Alpine; Dr. Glenn Storrs at the Cincinnati Museum of Natural History; and Amanda Masterson in Publications at the Texas Bureau of Economic Geology—all shared their expertise and experience.

Thanks to Phillip Ober and Expert Imaging, Inc., of Dallas, for expert film development and the enlarging of my photos. Additional thanks to Colorfast of DeSoto, Texas. The photos in this book are the author's work, unless noted.

William Lowe, former Editor-in-Chief of Gulf Publishing Company, conceived this project. Joyce Alff, Gulf's Managing Editor, deserves much credit for her patience and direction in seeing it to its conclusion.

My wife Rosa Finsley and sister Suzi Gibbons are major parts of that "author's family" group that never gets credit enough for freeing up time and providing great proofreading and general support for a long, time-consuming project. It wouldn't have happened without them. In fact, Rosa told me to do it!

This book presents the story of dinosaurs in Texas as I saw it, in my years of rambling from fossil lab to fossil lab across the state. It is a story of people as well as dinosaurs.

It was my sincere pleasure to bring it all together.

Preface

What does this book have to offer regarding Texas dinosaurs? While reviewing similar literature, I saw a real need for actual photos of Texas dinosaur fossils. Nothing can be closer to the creatures themselves than their fossils. Museum people feel that real things hold a special power. Fossils may be bent, broken, squashed, and twisted by eons of time while buried in the earth, but no drawing can capture entirely what you feel when you encounter the specimens themselves. Admittedly, photographs are not quite the real thing either; but I can't glue real fossils to these pages without getting in a lot of trouble.

I also felt there was a need for photos of the people who work with Texas dinosaur fossils, including a glimpse of their fascinating collection rooms, laboratories, and work benches. There is an exciting atmosphere where fossils are being studied. Such places are sometimes dusty, but not necessarily musty.

A very special feature of this book is that it uses all the dinosaur material in Texas, whether native to Texas or not. Some dinosaur fossils and replicas on exhibit in Texas museums are not from Texas. But, these are part of the Texas dinosaur experience. They illustrate parts of the Texas dinosaur story that actual Texas discoveries have yet to reveal; or they represent dinosaurs that we barely miss geographically in our Texas fossil record. This book shows you where these fossils are and tells you how they relate to Texas. In our discussion, we may cross state lines a little, when an important Arkansas, Oklahoma, or New Mexico reference adds to the Texas story.

Everyone who studies the subject of dinosaurs has a personal slant on the subject. Are modern fads in dinosaur interpretation realistic? How many of the older ideas about dinosaurs are still accepted? Where are the potential discoveries likely to be made in the future? Pulling this all together requires a little "art" among the science. That is my role, as a museum-oriented, earth-science-curating generalist who has dabbled in all kinds of fossils.

The story of dinosaurs in Texas is essentially a twentieth-century story, and this may be the last book on the subject this millenium.

There have been many scientific papers and at least two previous books on the subject of Texas dinosaurs. My good friends Tom and Jane Allen published *Dinosaur Days in Texas* in 1989, for which I penned a foreword. It is a picture book of the Texas dinosaurs that were known at the time, with a distinctly teacher's approach. I did my student teaching at Bryan Adams High School in Dallas, in a room next to Tom Allen's Texas History classroom. He considered everything, from dinosaurs to the Alamo, fair game for Texas history, which was his usual subject, and he taught every day with fire and brimstone. I remember him telling me about dinosaurs, "It's fascinating, just plain fascinating, that things like that lived here! Stand up there in front of the class and show 'em how ferocious a Tyrannosaurus could be!" Then he would actually shake and roar. He taught with passion. With Tom Allen's recent passing, we lost a great tour guide for Jurassic Park.

Dr. Louis Jacobs, of Southern Methodist University, wrote *Lone Star Dinosaurs* in 1995. The book was linked to a popular, traveling museum exhibit of the same name, both magnificently illustrated by Karen Carr. Dr. Jacobs benefitted greatly by his former association with Dr. Edwin H. Colbert, who was a long-time curator of paleontology at the American Museum. The Jacobs (Louis and his wife Bonnie) and the Colberts are close personal friends, and the book contains a wealth of stories about the American Museum and paleontologist Roland T. Bird. Bird discovered the well-known Glen Rose, Texas, dinosaur footprints. Of course Roland T. Bird's own fascinating book, *Bones for Barnum Brown,* (V. Theodore Schreiber, editor) is very much a book about Texas dinosaurs as well.

The discovery of dinosaurs shows us that the past holds mysteries. I don't think humanity will be happy on this little planet, if we ever solve the last mystery. We need mysteries to pursue. Scientists are just beginning to understand the mysteries associated with the dinosaurs. In Texas especially, the study of dinosaurs is in its infancy and still holds many mysteries.

The twentieth century began with very little being known about Texas dinosaurs. The next millenium will certainly hold fabulous discoveries for future Texas fossil hunters.

Getting Started

Beginnings

As a museum curator for more than 30 years, I was fortunate to move freely about a scientific community that studied dinosaurs. My purpose in writing this book is to give you a sense of what the people, the laboratories, and the dinosaur bones themselves are like.

First off, I want you to know how some of us got lucky enough to work with ancient fossil bones in the science of *paleontology.*

One of the most exciting books I have read is *Under a Lucky Star, A Lifetime of Adventure,* by world-renowned explorer, Roy Chapman Andrews. Later in life, Andrews became Director of the American Museum of Natural History in New York. The book is about his life as a naturalist, including his search for dinosaur fossils against the spectacular background of the Flaming Cliffs of Mongolia. His expeditions discovered the first known nests and eggs of dinosaurs, and many other fossil treasures. How does anyone get lucky enough to be such a person?

Andrews writes that when he first saw the American Museum of Natural History, he was overwhelmingly determined to work there, even if he had to sweep the floors! And he began by sweeping the floors. As opportunities came to him, he answered, "I can do that." This first led him to working with modern whale skeletons and behavior, then on to excavating dinosaurs, and finally traveling to far-off and romantic places.

When I was twelve years old, I made a visit alone on the city bus to the Clark County Historical Society Museum,

Dr. Louis L. Jacobs, Director of the Shuler Museum of Paleontology at Southern Methodist University (right) and preparator Andrew Konnerth are shown with fossils of the Texas sauropod Pleurocoelus. *A plaster jacket from that Hood County site weighed 11 tons.*

Robert (Bob) Rainey, chief preparator at the UT Austin Lab, is shown with a large brow horn of a horned dinosaur he found in the Texas Big Bend. The likely identification is Torosaurus *because* Triceratops *is not positively identified in Texas.*

in my hometown of Springfield, Ohio. The purpose was a school project of some kind. It was a dusty old place where, among a myriad of other things, there were Indian relics and a few modern and fossil bones. An old business card announced that the museum possessed, "Quaint and Curious Petrifactions." I still love that honest wording.

On a contemplatively slow bus ride home, I thought about the experience of those old exhibit cases and their

Author Finsley on the Heath mosasaur site, Rockwall County, Texas. (Photo by Douglas Hall.) See page 49 for a picture of the mosasaur on exhibit.

treasures. I determined right then, fully five years before I got my first museum job (sweeping floors), that some day I would help find wondrous things that people would want to put in museums and enjoy forever. I got my wish, many years later, while working with the fossil creatures that now grace the halls of the Dallas Museum of Natural History: a prehistoric elephant (mammoth), a swimming sea-lizard (mosasaur), a giant sea turtle, and a dinosaur. Most of these I dealt with from the hole in the ground, where they were found, to the exhibit floor, where visitors can enjoy them — in the museum version of forever. I found Roy Chapman Andrews' lucky star myself. If he and I were to meet, however, I think we would first discuss how many glass cases we polished and how many floors we swept.

The best such paleontological "how we got into this" story comes from the early experiences of my consultant on this book, Dr. Wann Langston, Jr. Few paleontologists have had as illustrious a career as Wann Langston. He is highly respected the world over for his fossil discoveries of Permian amphibians, extinct crocodilians, pterosaurs, and dinosaurs. I convinced him to put in his own words how he first got his hands on a real dinosaur.

People often ask professional paleontologists, "How did you first become excited about dinosaurs?" There is a common thread in the answer; many of us were bitten by the dinosaur bug at an early age. We saw a movie, our parents bought us a book, we saw dinosaurs in a museum, we made clay models of dinosaurs in school, and on and on. Most people outgrow dinosaurs (figuratively speaking)

the thousands of visitors who passed through the hall. When they had chiseled a bone out of the rock and mended it, they attached it to a wall in the appropriate place on a life-sized picture of a *Camarasaurus*. Thus, visitors could watch the animal grow.

When I walked into that building and spotted those bones, I never saw anything else at the exhibition. I stood in front of the barricade that separated the technicians and their dinosaur from the public, shifting one foot to the other and clearing my throat in hope of attracting some attention. The technicians ignored me. After a while, I could stand it no longer, and over the barricade I stepped, entering another world. I also got the technicians' attention. One of them, whose name was Norman Boss, headed in my direction saying something about the public not being allowed within his enclosure. He was obviously intent on ushering me out, when I asked him some technical-sounding question about *Camarasaurus*. I don't recall what it was, but I do recall he did not know the answer. He stopped and looked at me, and I threw another question at him, and he backed away. We struck up a conversation, he invited me to *hold* a piece of the bone. That was the closest I had ever been to a real dinosaur. He explained what they were doing and demonstrated the use of their tools and materials. I had read about all this in the few books that were available in the 1930s, but now I was actually experiencing how fossil bones are prepared for study and exhibition. I learned a lot in that wonderful hour or so. Many years later, when I was assembling dinosaur skeletons in Canada and Texas, I remembered those lessons and had the technicians prepare their fossils in public view, and mounted the bones on a diagram on a wall, just as I had seen it done in Dallas many years before.

I suspect Mr. Boss and his associate were glad to see five o'clock and quitting time roll around, but I would have happily stayed all night.

Dinosaurs and Texas

Texans are supposed to boast about having the biggest and best of everything. Yet, in terms of dinosaurs, Texas, indeed, has a *lot,* but not quite *everything*.

When most people think of dinosaurs, they think of great piles of bones found in the far western United States: Wyoming, Colorado, Utah, etc. Or they think of great museums in the older, eastern U.S. cities: New York, Washington D.C., or even midwestern Chicago. Until recently, most popular books on dinosaurs came from those same faraway places. When it comes to dinosaurs, many Texans are uncertain about where Texas stands in that regard. Just what has been found here?

In my 34-year career at the Dallas Museum of Natural History, I was involved in digging up many fossils: thousands of seashell creatures, several prehistoric elephants

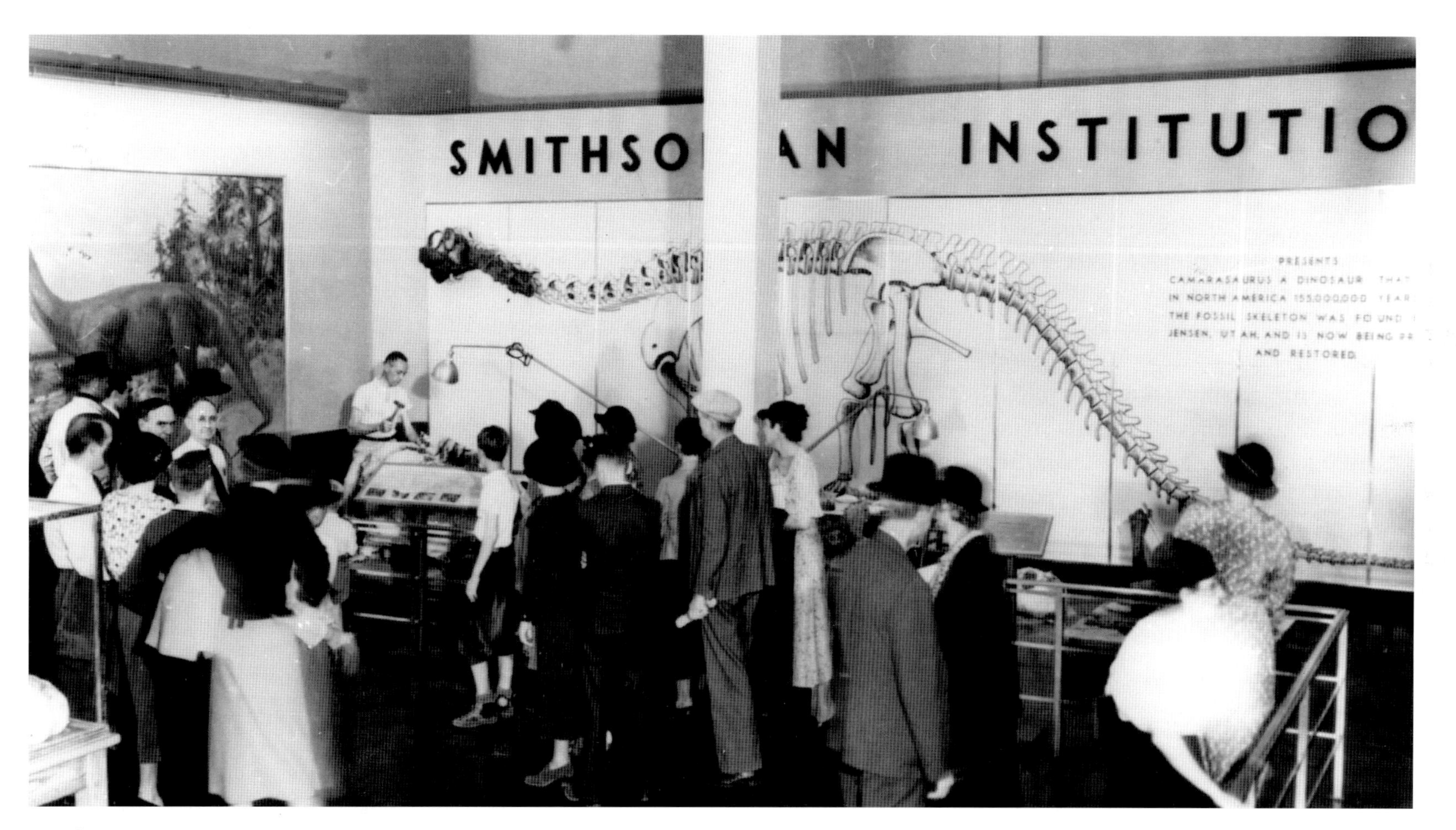

Smithsonian Institution's Camarasaurus *exhibit at the Texas Centennial Exposition in 1936.* Photo courtesy of the Office of Smithsonian Institution's Archives, Washington, D.C.

(mammoths and mastodons), several swimming sea reptiles (mosasaurs and plesiosaurs), and dinosaur-age fish galore, but only *one* fairly complete dinosaur.

For several reasons dinosaurs are not common Texas fossils. Dinosaur fossils occur in merely three of the dozen or so geologic periods. Smaller animals than dinosaurs occurred in much larger numbers. Dinosaurs were land animals and often died in situations more conducive to open-air decay than to preservation by burial in marine sediment. But, hard as dinosaur fossils are to find, the interest in dinosaurs is great, and year after year new discoveries expand the inventory of known Texas dinosaur species.

Discovering dinosaurs takes patience and a little bit of luck. A Texas dinosaur hunter may make only a few memorable discoveries in a lifetime, while fossil seashell collectors reap buckets of good material on every field trip. But who wouldn't want the pride and wonder of having found even one dinosaur bone?

Early Texas Geology

For quite a while after dinosaur fossils became known in many other places in the U.S., Texas remained off the well-beaten path of the famous dinosaur hunters of the 1800s. These hunters moved along the trails pioneered by the Gold Rush and the railroads, mostly north of Texas. Advanced university science programs and natural history museums barely existed in Texas in the nineteenth century. The Strecker Museum at Baylor University in Waco (1893) was the first real Texas museum, and natural history was just one of the subjects it covered.

The first interest in geology in Texas was economic geology: rocks and minerals, oil, and the occasional lost silver mine. These subjects were connected with a dollar value.

Texas geology really "welled-up" from the search for oil. That search was most interested in determining the age of the rocks, because it was found that oil was more likely in certain kinds and ages of rock. Dinosaur fossils, like most vertebrate fossils, are not very useful in such ways because they are too large and scattered to find in a small oil well drill hole.

Dinosaur fossils occur in relatively few beds of rock and are not widespread enough to help in matching up and identifying rock layers. Smaller, invertebrate fossils are abundant and useful in determining the age of most rock layers. Hence, Texas' early paleontologists concentrated on invertebrate fossils. The spiraled ramshorn fossils, called ammonites, were especially collected in the early days of Texas paleontology to determine the relative age of rocks.

Texas ammonite Drakeoceras. *Ammonites and other invertebrate fossils are widespread and plentiful enough to be more useful than vertebrate fossils for determining the relative age of rock layers.* Collection of John Moody, Sr.

Fossil fish vertebrae. Fish fossils (teeth and bones) are the most common vertebrate fossils in the marine Cretaceous rocks of Texas.

Many early Texas geologists, W. S. Adkins, W. M. Winton, and others were associated with invertebrate fossil work.

Texas rocks abound in invertebrate fossils such as corals, clams, snails, and ammonites. *Invertebrates* do not have a bony internal skeleton; specifically, they lack a backbone. Having less internal support, invertebrate creatures are generally smaller, but are more numerous than us, *vertebrate* creatures, with our internal skeletal support systems. Both invertebrates and vertebrates are descended from some unrecognized common ancestor. However, invertebrate animals come from many, many lineages; whereas, vertebrate creatures toe a much narrower line of descent: primitive fish-like animals without jaws or fins, followed by fish, then amphibians, reptiles, birds, and mammals. Because invertebrates are plentiful, widely found, and possess many distinct species adapted to many varied environments, they are much more useful to geologists in determining the age of rocks than are dinosaur bones.

Early discoveries of vertebrate fossils in North and Central Texas were generally fossil fish and sea-dwelling reptiles, such as the mosasaurs and plesiosaurs. They had the advantage of being buried in the same beds with the often-collected fossil seashells. Such creatures had all lived in a watery environment.

Marine (sea-bottom) rocks from before, during, and after the Age of Dinosaurs cover most of Texas. Land- or shoreline-deposited rock from precisely the Age of Dinosaurs is far less common. Such rock, capable of containing dinosaur fossils, is only well-exposed in the Texas Panhandle, parts of North and Central Texas, and the Big Bend region.

The great abundance of sea-deposited rock in Texas makes it difficult to find rock layers suitable both in age and dry-land environment that contain dinosaur fossils. In addition, land-deposited rock layers do not allow for rapid burial and protection of fossil bones. Of the millions of dinosaurs that once lived, the few that weren't eaten by predators or scavengers faded away into unrecognizable dust, their bones and teeth cracked by heat or cold, oxidized by the air, and dissolved by the rain. It is not surprising that early Texas geologists learned a lot about many other extinct fossil creatures before they ever saw a dinosaur bone.

Dinosaur fossil discoveries have always been a wonderfully special event. Texas dinosaur fossils are uncommon finds. That is why newspapers today love a story about someone finding a dinosaur. It is part of the public excitement about dinosaurs. "A thing like that lived here?" is a common reaction.

Any suspected dinosaur discovery by the public should be brought to the immediate attention of professional paleontologists at universities and museums. Some Texas dinosaurs are so rare they are known from only one or two tiny bone fragments or a few microscopic-sized teeth, as the photos in this book will show.

A Gallery of Texas Dinosaurs

This chapter illustrates and describes all the dinosaur species naturally found so far in Texas. Such a list is growing each year, and agreement among paleontologists is greater as to some species and less regarding others.

The Texas Triassic Dinosaurs

In Texas, we find rocks of the last part of the Triassic Period (the Dockum Group). Such rock layers are about 200 million years old. True dinosaurs had evolved by then, but they were relatively small animals, still competing for food and habitat with many other kinds of reptiles. Because these are the oldest dinosaur-bearing rocks in Texas, the small, scattered fossils of these oldest Texas dinosaurs are very important, but poorly preserved, and open to much speculation among paleontologists. Our gallery contains five possibilities from the last part of the Triassic Period, some more and some less agreed upon. These are *Coelophysis, Chindesaurus, Technosaurus, Shuvosaurus,* and *Tecovasaurus.*

The Texas Early Cretaceous Dinosaurs

There are extremely few surface rocks exposed in Texas from the Jurassic Period. Therefore, the next productive dinosaur-bearing rock layers after the Triassic Period are in the early to middle part of the Cretaceous Period. This was the part of the Cretaceous Period before a shallow Cretaceous sea covered most of Texas. Often during this period, the Gulf shoreline encroached northwest-ward over much of east and south Texas. Tidal flats occupied much of the present Texas Hill Country, with only slightly higher elevations to the north and west. This occurred about 115 million years ago.

Our gallery contains six possibilities from the early Cretaceous Period: *Acrocanthosaurus, Pleurocoelus, Iguanodon, Tenantosaurus, Deinonychus,* and a small hypsilophdont.

The Texas Mid to Late Cretaceous Dinosaurs

These dinosaurs range in age from 105 to 65 million years ago and occupied parts of Texas either during brief retreats of the vast, shallow Cretaceous sea or on shorelines exposed by its eventual retreat from much of Texas at the end of the Cretaceous Period.

Our gallery contains thirteen possibilities from this period: *Protohadro, Kritosaurus, Edmontosaurus, Pawpawsaurus, Panoplosaurus, Euplocephalus, Stegoceras, Ornithomimus, Chasmosaurus, Torosaurus, Alamosaurus, Tyrannosaurus,* and *Albertosaurus.*

Coelophysis

It is yet to be proven that the small, sharp, saw-edged teeth and scraps of bone so far found in Texas are those of *Coelophysis.* Fossil remains of this dinosaur have been found in rocks of similar age in New Mexico (see page 70 for photo). If the small teeth that have been found are any indication, then some member of the small, graceful, meat-eating coelurosaur family to which *Coelophysis* belongs was certainly in Texas. *Coelophysis* was a very small, lightweight, fast moving meat-eater, barely 6–8 feet long. Most of that length was a long, thin tail. Its hands were three-toed and capable of grasping prey. Its locomotion was primarily on its hind legs (bipedal). It may have hunted in packs, the way hyenas do today.

Chindesaurus

By comparing known fossils of *Chindesaurus* found in Arizona with fossils found in Texas, it can be said that *Chindesaurus* was definitely in Texas. Nonetheless, its overall anatomy is in need of better understanding by further discoveries. It appears closely related to one of the earliest known meat-eaters (theropods) on earth, the South American *Herrarasaurus*. *Chindesaurus* was 10 feet long, weighed about 200 pounds and was a large predator for Triassic times. It is known in Texas by such fragmentary evidence as a small femur head (see photo on page 120) and part of a hip (pelvis) bone.

Technosaurus smalli

Known from a small number of bones, including some herbivorous teeth, Dr. Sankar Chatterjee, who excavated the material, feels *Technosaurus smalli* was an ornithischian dinosaur. It has also been tentatively assigned to the prosauropods, a group of early plant-eating dinosaurs. *Technosaurus* was named for Texas Tech University, where Dr. Chatterjee teaches, and because it was found nearby in the Texas Panhandle. So few fossils of these small Triassic plant-eaters have been found that we are forced to make our assumptions from more plentiful African Triassic dinosaurs like *Lesothosaurus*, in the fabrosaur dinosaur family. Apparently they were 5 feet long and weighed only 60–70 pounds.

Shuvosaurus inexpectatus

This dinosaur was named for Dr. Chatterjee's son Shuvo and for the unexpected nature of its bones (seemingly too evolved for the Triassic Period). *Shuvosaurus'* large eye sockets and toothless beak would likely place it with supposedly much later bird-mimic dinosaurs of the Cretaceous Period, which would indicate millions of years of evolution beyond other Triassic dinosaurs. Overall paleontologic opinion is greatly divided on how to explain this specimen. Similar evolution among the non-dinosaurian rauisuchians of the Triassic, competitors of true dinosaurs, has been mentioned. Others say it looks very dinosaurian. (See photo on page 119.) *Shuvosaurus* was about 7 feet long and weighed approximately 100 pounds.

Tecovasaurus murryi

Dr. Phillip Murry, Tarleton State University, discovered a series of almost microscopic dinosaur teeth while working on his SMU doctoral dissertation about the Texas Triassic. *Tecovasaurus* was apparently a very small, plant-eating, bird-hipped dinosaur similar to the better known African dinosaurs, *Lesothosaurus* or *Fabrosaurus*. It was only about 3 feet long and weighed 35 pounds. Someday a matching skeleton will be found. So far *Tecovasaurus* is known by its distinctive teeth (see photo on page 121).

Acrocanthosaurus

Acrocanthosaurus is, at present, the one recognized big meat-eater of the Early Cretaceous Period in Texas, 110–120 million years ago. *Acrocanthosaurus* preceded the tyrannosaurs by almost 50 million years. Earlier Jurassic Period meat-eaters, such as *Allosaurus* (see photos on page 75), may have lived in Texas, but a lack of Jurassic surface rock exposures makes that presently unprovable. *Acrocanthosaurus* was 35 feet long and weighed 12,000 pounds. Its backbone has a ridge of moderately high spines that were anchor points for strong back muscles. Perhaps, by seizing its prey in its dagger-like teeth, it used a violent side-to-side thrashing to finish the job (see photos on pages 81, 107, 108, 131, and 137).

Pleurocoelus

The brontosaur-like dinosaur of the Texas Early Cretaceous was *Pleurocoelus*. The earlier Jurassic "brontosaurs," like *Diplodocus* and *Apatosaurus* (see photos on pages 73 and 74) are as yet undiscovered in Texas' poorly exposed Jurassic rocks. *Pleurocoelus* was in the brachiosaur family. Great herds of up to one hundred of these huge beasts roamed Arkansas, Oklahoma, and Texas 110–120 million years ago. Big as it was, 50 feet in length and weighing 40 tons, *Pleurocoelus* was only Texas' second largest dinosaur, compared to the later and larger *Alamosaurus*. *Pleurocoelus* is thought to have made the bathtub-sized footprints at Glen Rose, Texas, and elsewhere in early Cretaceous rock (see photos on pages 81, 109, 131, 136, and 137).

Iguanodon

This is probably the largest ornithopod dinosaur to live in Texas. *Iguanodon* could be 30 feet long and weigh over 5 tons. *Iguanodon* remains are infrequent compared to big ornithopods like *Tenontosaurus* or the much later duckbills. It is famous for the sharp spiked thumbs on its forelimbs. Many fat, three-toed footprints have been tentatively identified as those of *Iguanodon,* but only fragmentary bones have been discovered in Texas (see photo on page 82).

Tenontosaurus

The most commonly found large plant-eating ornithopod dinosaur in Texas during the Early Cretaceous, *Tenontosaurus* was 20 feet long and weighed about 1.5 tons. Its name refers to the plentiful stiff tendons that supported its tail, making it clear that *Tenontosaurus* was not a tail-dragger (see photos on pages 101–107 and 114).

Proctor Lake Ornithopod Dinosaurs (Hypsilophodonts?)

This is an example from a large group of fossil skeletons of medium to small plant-eating dinosaurs found near Stephenville, Texas. These dinosaurs were either in the hypsilophodont family (also the family of the *Tenontosaurus)* or closely related. The larger Proctor Lake specimens are 6–8 feet long and weigh only 75–125 pounds. They were obviously a herd animal, and judging from the large number of skeletons found in a limited area, Proctor Lake was possibly a nesting ground. No eggs are presently known, although very juvenile specimens were in the grouping (see photos on pages 109, 110, and 111).

Deinonychus

Evidence for the existence of *Deinonychus* in Texas is debatable. Further north, *Deinonychus* was known to feed on dinosaurs like *Tenontosaurus,* which do occur plentifully in Texas Early Cretaceous rocks. Its name and trademark was the "terrible claw" on the first toe of its hind foot. The claw acted as a "can-opener" on its prey. *Deinonychus* was approximately 12 feet long and weighed 750 pounds. A tooth from a dromaeosaur, the same family as *Deinonychus,* is shown in the photo on page 86.

Protohadro byrdi

This is one of the earliest duck-billed dinosaurs discovered anywhere in the world. It lived in northeast Texas, just north of what is now the DFW Metroplex, about 95 million years ago. It was found by and named after Gary Byrd, a very diligent Texas fossil collector. It was a very basic, modest-sized member of the hadrosaur family. *Protohadro* was about 25 feet long and weighed almost 2 tons (see photos on page 126).

Kritosaurus navajovius

The Aguja Formation in the Texas Big Bend contains fossils of a duck-billed dinosaur whose wide nose bends slightly downward. Such dinosaurs have been called *Kritosaurus*. It was 25–30 feet in length and weighed 1–2 tons. This was the earliest duck-billed dinosaur of the Texas Big Bend, although duck-billed dinosaurs lived in northeast Texas perhaps 20 million years earlier (see photos on pages 87, 96, and 112). Shown here also is the fossil crocodile, *Deinosuchus*, a Texas dinosaur-age creature big enough to have killed and eaten all but the largest dinosaurs. It reached lengths of 30–40 feet and lurked in Texas Big Bend waterways.

Edmontosaurus

Somewhat sketchy evidence found in the Javelina Formation, the latest sediments of the Big Bend area of Texas, indicates the existence of a moderately large, crestless duck-billed dinosaur. Similar remains found in other places in North America are from *Edmontosaurus.* It ranged in size upwards from 30 feet in length and could weigh 2–3 tons. This dinosaur takes its name from Edmonton, Canada. It is likely the Texas genus and species from the Javelina Formation will be named differently when evidence is better known (see photo on pages 40 and 74).

Pawpawsaurus campbelli

Pawpawsaurus was an armored dinosaur in the nodosaur group. Nodosaurs possessed side spines, but lacked the tail clubs and very broad heads of the similarly armored ankylosaur group. It was approximately 15 feet long and weighed 3,000 pounds. It was found just north of what is now the DFW Metroplex in sediments of a brief retreat of the sea in North Texas about 100 million years ago. It was a time when most of the rest of Texas was still under the shallow Cretaceous sea (see photos on pages 127 and 128).

Panoplosaurus

Panoplosaurus was an armored nodosaur primarily from the Big Bend Aguja Formation. It was the last North American nodosaur and probably continued in Texas until the end of dinosaur times. *Panoplosaurus* was over 20 feet long and could weigh well over 2 tons.

Euoplocephalus

The ankylosaur *Euoplocephalus* is best known from the Big Bend Aguja Formation, but fragments of some ankylosaurs occur also in the later Javelina Formation. It was (like the nodosaurs that were also found in West Texas) over 20 feet long and weighed over 2 tons. These were armored giants on the edge of the then retreating Cretaceous sea. The nodosaurs and ankylosaurs needed their weight and armored skin to resist large predators, which were either *Tyrannosaurus* or something very, very similar.

Stegoceras

This thickly boneheaded dinosaur (a kind of pachycephalosaur) is found in the Aguja Formation of the Texas Big Bend. It was rather small, only 6 feet long and weighing 100 pounds. Its diet was probably plants and insects. The thick skull was most suited for settling disputes with other Texas "boneheads."

Tyrannosaurus and *Ornithomimus*

Tyrannosaurs could reach 30–40 feet in length and weigh 5–7 tons. Large tyrannosaurid teeth and bones have been found in the Javelina Formation in the Texas Big Bend. Measurements of these incomplete remains cast some doubt on whether they are precisely *rex* itself. That may mean that we have our own kind of *Tyrannosaurus* in Texas (see photos on pages 41, 85, and 97).

Ornithomimus, a Late Cretaceous bird-mimic dinosaur, is also shown here. These were very fast-moving, agile, small dinosaurs. Toothless, like modern birds, ornithomimids probably ate a varied diet of plants, insects, eggs, and small animals. They are poorly known from the Texas Big Bend, primarily from partially identified remains in the Aguja Formation. If, as our artist depicts, they ever co-existed with *Tyrannosaurus,* could the much bulkier king of dinosaurs have caught the fleet-footed *Ornithomimus?* Certainly not often!

Chasmosaurus mariscalensis

This horned dinosaur with a large, wide frill of bone above the eyes occurred only in the older Aguja beds of the Big Bend. *Chasmosaurus'* skull frill has several openings across it (think of the word "chasm," meaning big hole). It had the usual ceratopsian family horns on the skull, but the horns of *Chasmosaurus* were quite varied—some short and stocky, some long and flaring. Sex differences have been suggested, with male horns perhaps more protectively straightforward for defense. This was a moderately small horned dinosaur, about 15 feet in length and weighing about 2 tons (see photo on page 113).

Torosaurus

This large horned dinosaur, with a particularly long frill of bone above the eyes, occurred in the later Big Bend sediments, the Javelina Formation. *Torosaurus* is the stand-in in the Big Bend of Texas for the more famous *Triceratops*, which is not yet certainly identified in Texas. It was 25 feet long and weighed approximately 10,000 pounds. The University of Texas has some isolated horns and bones that could be either *Torosaurus* or *Triceratops* (see photos on pages 2, 43, and 88).

Alamosaurus

This huge member of the titanosaur family, primarily a south-of-the-equator group, wandered north into the Texas Big Bend at the end of the dinosaur age. South America had been detached from North America and must have just drifted back into contact via Central America or a series of islands. *Alamosaurus* was Texas' largest dinosaur—approximately 70 feet long and weighing 60,000 pounds. Titanosaurs can sport an armored skin, which we hint at in this illustration. Many disarticulated and jumbled alamosaur fossils are found in the Javelina Formation in the Texas Big Bend.

Alamosaurus did not get its name from any connection with the famous Texas mission in San Antonio, Texas. It was in fact named for a New Mexican spring "*ojo*" that watered a cottonwood tree "*alamo,*" a familiar combination in the arid southwest. The first discovery of *Alamosaurus* fossils was made near there (see photos on pages 4, 38, 39, 87, 100, 122, and 123).

Albertosaurus

Bones and teeth of tyrannosaurid meat-eaters occur in the Javelina and Aguja formations of the Texas Big Bend. The largest of these remains, especially those with large teeth, are generally called *Tyrannosaurus*. Such occur especially in the Javelina Formation. Smaller remains in both formations are called "something-like-an" *Albertosaurus*. *Albertosaurs* were perhaps 30 feet in length, yet had all the scary habits associated with *Tyrannosaurus*. *Albertosaurs* were probably even faster moving and more agile. Classic *Tyrannosaurs* and *Albertosaurs* were northern U.S. and Canadian creatures. It may be that all of these tryannosaurid fossils in Texas will be recognized as two or more new genera and species when additional discoveries are made. (See *Albertosaur* skull drawing, page 40.)

Some Dinosaur Basics

Of Ages and Eras

Dinosaur times, the so-called Age of Dinosaurs, was just one particular segment of earth's history. Scientists call it the Mesozoic Era meaning "middle life." It is a period of time between the two largest extinctions of life on earth. It was a good time for the emergence of wonderful, and very different, things.

Earth history is divided into extremely large chunks of time, called "eras," based on our understanding of how earth history was changed by some very large events. The length of time between such great events we call individual eras. It is just an imperfect, human way of trying to make sense out of nearly five billion years of the earth's past.

Time *before any* life can be considered one early era, the **Archaen** or **Archaeozoic** Era. The period with the earliest beginnings of life on earth and *with very simple life-forms* can be a second era, sometimes called the **Proterozoic Era.** About 550 million years ago a rather sudden and *great increase of variety in life-forms* occurred. That marked the beginning of a third era, the **Paleozoic Era.** The development of life on earth made great progress during the Paleozoic Era. Invertebrate life-forms gave rise to early vertebrate forms, the precursors of fish. Fish gave rise to vertebrate land creatures such as amphibians and reptiles. Reptiles during the Paleozoic Era diversified into different and ever more specialized forms. Mammals, like ourselves, trace our development back to the reptiles of the late Paleozoic Era.

Then came earth's greatest extinction. At the end of the Paleozoic Era, perhaps 90 percent of earth's life-forms

Geologic Time Chart

ERAS	TIME (MILLIONS OF YEARS AGO)
Cenozoic	65–present
Mesozoic	245–65
Paleozoic	570–245
Proterozoic	2500–570
Archaeozoic	4600–2500

vanished. It was with that background, the loss of many kinds of living things, that the next era, the Mesozoic Era (Age of Dinosaurs), dawned. The Mesozoic Era, itself, would later end in another time of wide extinction that wiped out the last dinosaurs and many, many other lineages of living things. That second large extinction came at approximately the rise of the present era, the Cenozoic Era (the era of recent life forms), also termed, the Age of Mammals.

In the future, scientists (perhaps, you!) will tear this old fabric of neat eras apart and build a new and better scheme. But, this is the way it looks from the end of the twentieth century.

Remember that some particular species of animals and plants are in the process of becoming extinct all the time. Seemingly related extinctions of a large number of species have occurred every few million years. The two extinctions at the dividing points of the Paleozoic and Mesozoic eras, and later the Mesozoic and the Cenozoic eras, were merely two especially big events. In fact, small extinctions from a variety of environmental and biological causes are fairly common and natural as earth's conditions change.

Paleozoic reptiles had been around for about a hundred million years, before some reptiles changed (evolved) enough to become what we call dinosaurs. If one views the complete history of the earth, dinosaurs are a rather recent development. They appeared about 230 million years ago and became extinct only about 65 million years ago, out of a stream of life that is billions of years old.

So the earth is much, much older than the dinosaurs, and some living things were around a lot longer than the dinosaurs. As earth history goes, dinosaurs belong to our modern end of it.

First Hints of Dinosaurs

Dinosaurs' ancient existence was first realized only about 175 years ago. Between 1822 and 1840 suspiciously unique teeth and bones, many of the dinosaur *Iguanodon,* began to turn up in England and in Europe. Simultaneously, huge bird-like tracks were noted in Triassic sandstones of New England. Such remains didn't even have the formal name, dinosaur, until 1841. An early English paleontologist, Richard Owen, finally, officially coined the term. Not that lack of names had bothered even one dead dinosaur!

These bones popped up as a surprise even to scientists. That surprise may be one of the most important things about dinosaurs. Such surprises remind us that the past can contain immense new concepts, truths, and even ten-ton creatures! There is much about the earth and its inhabitants, past and present, we do not yet know. Dinosaurs are fantastic reminders of that humbling notion. Once we know about dinosaurs, we cannot look at a hill or a rocky cliff quite the same way again. What wonders are buried there?

As more and more dinosaur fossils were found, it became clear that they were reptiles, yet quite unlike any living reptiles. The first clue to their uniqueness was that the first dis-

Dr. Marianne Levine, Dallas Paleontological Society, on a field trip to the UT Austin Vertebrate Paleontology Lab, with tail vertebrae from an immense Alamosaurus *sauropod dinosaur from the Texas Big Bend.*

covered dinosaur remains were of enormous size. That is still what most people think and feel about dinosaurs. It was much later that scientists realized that there were also smaller dinosaurs. Enormous size still vibrates in the common name *dinosaur.* If all dinosaurs had been small, they would not have been as quickly discovered, and most would be

Triceratops *skeleton replica on exhibit at the Witte Museum, San Antonio, Texas. Although several horned dinosaurs are clearly native to Texas,* Triceratops *has only been positively identified farther north.*

Texas Dinosaurs at a Glance

Saurischian (Lizard-Hipped) Dinosaurs

GROUP	NAME	PERIOD	DIET	SIZE	LOCATION
Theropods	Chindesaurus	Triassic	Carnivorous	10 ft. long 200 lbs.	West Texas and Panhandle
	Shuvosaurus	Triassic	Carnivorous	7 ft. long 100 lbs.	Texas Panhandle
	Coelophysis	Triassic	Carnivorous	7 ft. long 65 lbs.	Only suspected in Texas
	Acrocanthosaurus	Early Cretaceous	Carnivorous	35 ft. long 12,000 lbs.	N. Central Texas
	Deinonychus	Early Cretaceous	Carnivorous	12 ft. long 750 lbs.	Only suspected in Texas
	Albertosaurus	Late Cretaceous	Carnivorous	30 ft. long 10,000 lbs.	Texas Big Bend
	Tyrannosaurus	Late Cretaceous	Carnivorous	35 ft. long 13,000 lbs.	Texas Big Bend
	Ornithomimus	Late Cretaceous	Carnivorous	10 ft. long 200 lbs.	Texas Big Bend
Sauropods	Pleurocoelus	Early Cretaceous	Herbivorous	50 ft. long 80,000 lbs.	N. Central Texas
	Alamosaurus	Late Cretaceous	Herbivorous	70 ft. long 60,000 lbs.	Texas Big Bend

Ornithischian (Bird-Hipped) Dinosaurs

GROUP	NAME	PERIOD	DIET	SIZE	LOCATION
Ornithopods	Tecovasaurus	Triassic	Herbivorous and insects	3 ft. long 35 lbs.	Texas Panhandle
	Technosaurus	Triassic	Herbivorous	5 ft. long 70 lbs.	Texas Panhandle
	Iguanodon	Early Cretaceous	Herbivorous	30 ft. long 10,000 lbs.	West Texas, suspected elsewhere
	Tenontosaurus	Early Cretaceous	Herbivorous	20 ft. long 3,000 lbs.	N. Central Texas
	Hypsilophodontids	Early Cretaceous	Herbivorous	6 ft. long 85 lbs.	N. Central Texas
	Protohadro	Mid Cretaceous	Herbivorous	25 ft. long 4,000 lbs.	N. E. Texas
	Edmontosaurus	Late Cretaceous	Herbivorous	35 ft. long 5,500 lbs.	Texas Big Bend
	Kritosaurus	Late Cretaceous	Herbivorous	30 ft. long 4,500 lbs.	Texas Big Bend, possibly N. E. Texas
Ceratopsians	Torosaurus	Late Cretaceous	Herbivorous	25 ft. long 10,000 lbs.	Texas Big Bend
	Triceratops	Late Cretaceous	Herbivorous	30 ft. long 12,000 lbs.	Suspected in Texas Big Bend
	Chasmosaurus	Late Cretaceous	Herbivorous	16 ft. long 8,000 lbs.	Texas Big Bend

(table continued on page 46)

Ornithischian (Bird-Hipped) Dinosaurs (continued)

GROUP	NAME	PERIOD	DIET	SIZE	LOCATION
Others	Pawpawsaurus	Mid Cretaceous	Herbivorous	15 ft. long 3,000 lbs.	N. E. Texas
	Panoplosaurus	Late Cretaceous	Herbivorous	20 ft. long 6,000 lbs.	Texas Big Bend
	Euoplocephalus	Late Cretaceous	Herbivorous	20 ft. long 5,500 lbs.	Texas Big Bend
	Stegoceras	Late Cretaceous	Herbivorous	6 ft. long 120 lbs.	Texas Big Bend

(text continued from page 42)

Other Aspects of Dinosaur Study

Nonetheless, this is a very mechanical way to define a dinosaur by skull openings or by hip and thigh bones. Remember, however, that bare bones are mostly what we can study in paleontology. If soft parts (heart, lungs, etc.) were commonly preserved as fossils, scientists strongly suspect that dinosaurs would appear even more different from other reptiles. Scientists have long debated whether dinosaurs may have developed ways to stabilize their body temperature as mammals do today, so-called warm bloodedness. In the absence of preserved soft tissues, such speculation must rely on characteristics of bone and guesses about lifestyles. Sustained high activity levels are clues that an animal had some way of regulating its internal thermometer to put forth such energy without overheating. It would be nice to have a living dinosaur to study.

There Were *Almost* All Kinds of Dinosaurs

One of the most fascinating things about dinosaurs is how they evolved into so many different varieties: plant-eaters or meat-eaters, fast or plodding, thin or fleshy, large or small, scaly or feathery (?). They also evolved to live in so many different environments.

One very puzzling fact about dinosaur evolution is that no non-avian (non-bird) dinosaur adapted to a life in the ocean, as some reptiles such as ichthyosaurs, plesiosaurs, and mosasaurs did, or as whales have done during the Age of Mammals. Here is an interesting speculation by Dr. Glenn Storrs, curator of vertebrate paleontology and specialist on ancient marine reptiles at the Cincinnati Museum Center and an expert on ancient marine reptiles.

Dinosaurs as Marine Reptiles?

To the best of our knowledge, with the noteable exception of birds, the diverse group called Dinosauria never successfully exploited the marine realm. Birds are widely held to be dinosaur descendants, part of the dinosaurian radiation, and thus dinosaurs in their own right (avian theropods, if you will). Therefore, marine birds such as penguins can legitimately be considered "marine dinosaurs." With this exception, there is no evidence of ocean-going dinosaurs (although occasional carcasses drifted out to sea, later to be fossilized). Why should this have been so?

Three basic obstacles to the invasion of the ocean by dinosaurs can be postulated. Were dinosaurs excluded from this arena by competition from preexisting marine groups? Were they physiologically incapable of making the land to ocean transition? Could a more basic anatomical factor have prohibited their evolution in that direction?

The suggestion that competitive exclusion explains the dinosaurs' absence from the marine record presupposes that every potential ecological niche was filled by other organisms. Plesiosaurs, ichthyosaurs, and mosasaurs are just three of the many reptile lineages that had terrestrial forebearers but later entered the sea. They presumably were exploiting new ecological resources that gave them a selective advantage for a life at sea. Could not dinosaurs have done this? Were there too many reptiles in the sea by then? The trouble is that certain kinds of reptiles "returned to the sea" numerous times during the Mesozoic, even after dinosaurs were well established on land. It's true that plesiosaurs and ichthyosaurs got their starts before dinosaurs in the Triassic, but mosasaurs do not make their appearance until the Cretaceous. Fully marine crocodiles are seen in the Jurassic, when dinosaurs had long been flourishing alongside their terrestrial cousins. If there was "ecological space" for the crocodiles, mosasaurs and others, there should have been room for the dinosaurs. Indeed, we see avian dinosaurs (birds) in the Cretaceous seas.

Dinosaur physiology is a thorny issue, especially over the question of dinosaur "warm" vs. "cold-bloodedness." Without living non-avian dinosaurs, evidence regarding their metabolism is inferential and problematic. The current consensus is that they lay somewhere between living reptiles and mammals/birds. A wide variety of groups with different metabolic strategies have been, and are, marine. So why would the adjustment from land to sea have been impossible for dinosaurs? Here again, other groups have had little problem with the change. Probably the best argument against a physiological barrier to a dinosaurian invasion of the sea is that some of their own, presumably with rather similar physiologies, did get in. Remember those birds?

Just what did keep most dinosaurs on dry land? The possibility of anatomical constraint is intriguing. What is it that makes a dinosaur, after all? A major feature is the unique (among reptiles) upright stance or posture of the dinosaur limb. All other reptiles, including those that did become marine, have a more or less sprawling stance. Most of these marine reptiles adopted a sinuous undulation of the body and tail for swimming. The unique turtles and plesiosaurs were prevented from doing so by a stiff body or shell that forced them to use their limbs. Perhaps the upright stance of dinosaurs and their long, often stiff tails kept them from having a viable swimming style. Mammals, although having an upright stance, have a very different degree and style of flexibility in their spinal column. What about birds, you say? Birds are unique among the dinosaurs, and indeed among the tetrapods, in having two distinct and independent locomotory systems. Although they stand upright and have stiff bodies, birds have wings and can fly. As a matter of fact, penguins do nothing more than "fly" underwater. Other birds use strong paddle strokes of their feet for swimming but, like penguins, have lost the long bony tail of more primitive dinosaurs. Perhaps dinosaurs could only enter the sea after abandoning the ground for the air.

Although we can never really know, something about the dinosaurs, possibly their erect stance, kept them away from a life on the ocean waves. Maybe we should just forget about the absence of "marine dinosaurs" and think of sea birds when we fret over this quandry. In this very real sense, dinosaurs made it into the sea after all.

"You Old Dinosaur!"

Many people mistakenly call any large fossil reptile a "dinosaur." Some pitifully ill-informed people call the large fossil mammals, such as prehistoric elephants,

The Heath, Texas, mosasaur (Tylosaurus) *on exhibit, Dallas Museum of Natural History. This was a reptile that found a home in the sea during the Age of Dinosaurs. Note the sprawling legs of a non-dinosaurian reptile.*

dinosaurs. Men's clubs, where I often speak, call some of their members dinosaurs. Politicians call their opponents dinosaurs, hardly a compliment to the real thing! So "dinosaur" means a very different thing to scientists than it does to some people.

The "Fossil Farmer"

This discussion brings to mind Jimmy Joe Herndon, who was plowing a field just east of Dallas when he looked down from his tractor seat into a shallow gully and saw a jumble of fossil bones. Jimmy had seen fossil bones before while visiting the Dallas Museum as a school boy. We agreed to meet at the Dairy Queen in Royce City, Texas; then he proudly took me out to see his find. The fossils were, in fact, a wonderful discovery of a huge dinosaur-age sea turtle. We later reconstructed it in the Dallas Museum of Natural History. Jimmy's first reaction, as he later reported to the press, had been, "My gosh, I've found my own dinosaur!" Needless to say, as interesting as it was, it wasn't a dinosaur. It was an eight-foot wide *Protostega* turtle. It might come as a surprise to some people that I value that giant fossil turtle every bit as much as if it were a dinosaur.

Some Identification Hints

Here are some hints in identifying a fossil bone, track, or whatever as possibly from a dinosaur.

- Try to determine the geologic age of the rock that contained the bone. Even totally unintelligible scraps of fossil bone from Paleozoic rock or from bedrock well into the Cenozoic Era can be, by default, credited to something other than a dinosaur.
- Determine something about the ancient environment that the rock represents. Unless you want to speculate that some sliver of bone is from a dinosaur carcass floating far out to sea, disqualify any fragment of bone from a marine sediment (unless the object is clearly some part of known dinosaur anatomy).
- Finally, determine by shape whether it is a dinosaur bone or perhaps the look-a-like bones of other dinosaur-age animals.

There is no substitute for a good knowledge of anatomy. Paleontologists generally specialize in some particular part of fossil history. In that way, they can become extremely familiar with the anatomy of a smaller number

Dr. Melissa Winans, collection manager, Vertebrate Paleontology Lab, UT Austin, cataloging fragment of duck-billed dinosaur jaw found in marine rocks, Ozan Formation, N. Sulfur River, Fannin County, Texas. It was collected by Monte Payton for Dr. Joan Echols, East Texas State University, and later acquired by UT Austin.

An unlikely rock for dinosaur fossils, duck-billed (hadrosaur) teeth show clearly in this specimen from the Sulfur River in northeast Texas.

of creatures within their field of interest. That allows them to go well beyond dinosaur basics.

Fossil hunters sometimes find "fossil" objects, such as fossil excrement called *coprolites* and stomach stones called *gastroliths.* In order for such unusual things to be considered genuine, they must be found in close association with the fossil creatures from which they supposedly came. There are so many naturally occurring objects that

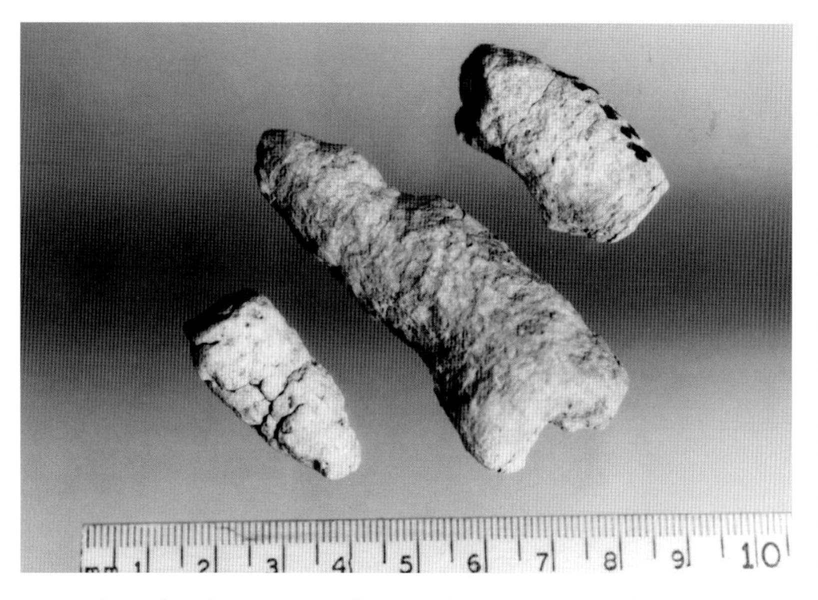

Coprolites, fossil excrement, from rock that contains dinosaur fossils, Texas Big Bend, Presidio County, Texas. These specimen may or may not be dinosaurian.

resemble gastroliths, for instance. This is another reason to seek professional advice *before* digging. Fossils wait millions of years in the ground to be discovered. A bad job of removing them can erase millions of years of important fossil information.

Gastroliths, stomach stones, from an Oklahoma rock formation containing dinosaur fossils. Unless found in actual skeletal remains (these were not), such fossils are purely speculation. Fun to imagine, but not science!

Is Your Chicken Thigh a Dinosaur Clue?

Some very exciting thinkers, among them Dr. Tim Rowe at the University of Texas at Austin, regard birds as up-to-

Dr. Timothy Rowe, director of the Vertebrate Paleontology Lab at UT Austin, with a fossil bird leg bone (femur) from the Texas Big Bend. Dr. Rowe considers such bones to be both dinosaur and bird (avian dinosaurs).

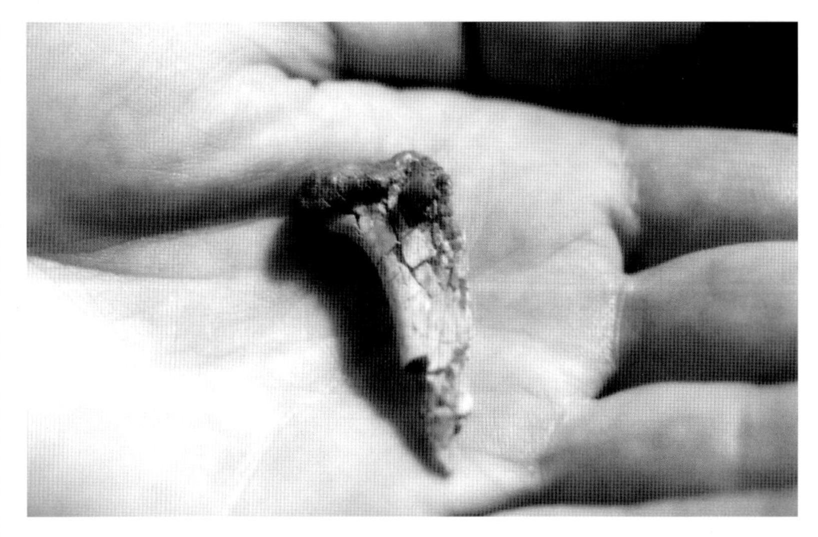

Note angled femur head of this fossil bird leg bone is reminiscent of the dinosaurs.

date dinosaurs. When I recently showed him a fossil bird bone from West Texas and asked him if it was bird or dinosaur, he replied, "Both!" By this way of thinking, dinosaurs are not only fairly modern, they are still here with us. Then we must re-think if dinosaurs did become, exactly and entirely, extinct. Drs. Lowell Lingus and Timothy Rowe have written a book on that subject entitled *The Mistaken Extinction.* Personally, I have some trouble taking much comfort over losing all those ancient and massive dinosaurs, just because we still have the birds. Compared to the massive brontosaurs, birds are, at most, only little comforts. But, Drs. Lingus and Rowe represent a coming trend that questions whether a dramatic, sudden dinosaur extinction isn't just an illusion caused by too few specimens and the inaccuracy of current rock dating techniques. It seems more and more likely, from the cutting edge of paleontology, that dinosaurs met a more normal and gradual demise, based on climatic changes and disease, perhaps helped along by an asteroid crash or two. Scientists should always remember that the really unusual rarely happens.

Life at the beginning of the Mesozoic Era would have looked like an interesting competition between varied reptilian groups in which the dinosaurs' ancestors and our distant relatives and many other reptilian groups were fairly equal players. For most of the early part of the Mesozoic Era, it wasn't the Dinosaur Age at all. It would have been difficult to pick a winner among the various competing reptile groups. But nature intervened, and at the end of the first geologic period in the Mesozoic Era (the Triassic Period), dinosaurs, crocodilians, and pterosaurs escaped an extinction that removed most of their reptilian competitors.

For the remainder of the era, the increasingly diverse kinds and numbers of dinosaurs gave them dominance over vertebrate life on earth. It might have seemed then that dinosaurian evolution was the sure direction life on earth would take from then on. Such would have been a logical view right up to the dinosaurs' final and unexpected extinction (except perhaps for the birds) at the end of the Mesozoic Era. At the time of the extinction of the last dinosaurs, there were no mammals larger than a house cat. Except that things always change, there seemed no necessity for a coming end for the dinosaurs.

Pesky Things!

Dr. Dale Russell, a well-known North American paleontologist from North Carolina State University, once speculated that given uninterrupted, continued evolution, some species of manually dextrous dinosaurs, using their "hands" to manipulate things and therefore open to a little more mental work than most reptiles, might have developed bigger, better brains and achieved that difficult-to-define condition called "intelligence." He explains a "pesky" advantage the dinosaurs might have possessed.

> Let us assume that a struggle for limited resources exists in nature, and those animals that are more "fit" are more apt to transmit their attributes to their successors. Modern invasive ("pest") animals are typically more active, have more complex central nervous systems, and originate in regions of greater biodiversity. One may suspect, therefore, that animals have become more aggressively "pest-like" through the effects of natural selection over even longer intervals of geologic time. Dinosaurs would thus be a scourge in a coal-age world, and had they survived to the present, dinosaurs would have become even greater "pests."

A Green Side of the Story

The study of fossil plants is called *paleobotany.* Plants are so basic to any study of life on earth that we can't neglect a mention of them in regard to dinosaurs. During the Age of Dinosaurs, several great changes in plant life occurred. The plant life of the Late Paleozoic (before the dinosaurs) had been dominated by what can very generally be called ferns (seed-ferns and pteridophytes). These had soft vegetation at a variety of heights off the ground. The principal herbivores were rather small creatures, adapted for browsing at no great height. At about the time of the first dinosaurs, a season-changing, drier, cooler climate replaced much of the ferns with conifers and cycads. These newly dominant plants had courser texture, less digestability, and were often (the conifers) on taller trees. This seems to have favored the life styles of long-necked herbivores, beginning in the Late Triassic with prosauropod dinosaurs like *Plateosaurus* and continuing through the Jurassic period with the famous long-necked, brontosaurian and brachiosaurian giant sauropods. These creatures relished using their peg-like teeth to strip the greenery off a high branch, but with those peg teeth, were rather unable to chew it thoroughly. The change in climate probably had to do with the coming together of earth's continents to form one large continent, which we shall discuss in the next chapter. Smaller herbivores were present, but their numbers did not "take off" until much later when the next great dinosaur-age plant revolution occurred.

That was the evolution and spread of flowering plants (angiosperms). The flowering plants evolved rather late in dinosaur times, the Late Cretaceous. That means all those popular Jurassic dinosaurs (*Apatosaurus, Diplodocus, Stegosaurus* and *Allosaurus)* never ever saw or smelled a beautiful flower.

Flowering plants included many fast-spreading species, able to quickly spread seeds on disturbed ground. We, in fact, call many of these "pioneer plants" because they are so able to propagate in a wide variety of environments. Some paleontologists feel the earlier giant "peg-toothed limb-rakers" in some places destroyed large areas of former conifer forest, making way for the smaller, faster-growing angiosperms. The world before the flowering plants had probably been a more barren landscape, in any event. Our modern experience of plentiful plant cover was not so much the case during the earlier two-thirds of the dinosaurs' world.

Once flowering plants with their lower, softer, more digestible vegetation became widespread, smaller, shorter,

A 95-million-year-old fossil leaf, Woodbine Formation, Denton County, Texas. Such broadleaf plants produced for the first time a more modern type of vegetation in the Late Cretaceous Period.

Dr. Bonnie Jacobs, a paleobotanist associated with the Shuler Museum of Paleontology at SMU with (rear) Cycadeoidea, *a Texas cycad fossil (Early Cretaceous) and (front) Texas broadleaf plant fossils (Late Cretaceous) from later in the dinosaur age. These fossils illustrate vegetational change in the dinosaur age.*

plant-eating dinosaurs, especially the horned dinosaurs (ceratopsians) and the duck-billed dinosaurs (hadrosaurs) rapidly expanded their numbers. Dr. Tom Lehman, of Texas Tech University, studied the distribution of plant-eating dinosaurs in the Texas Big Bend, and found evidence that the last of the great long-necked sauropod dinosaurs, *Alamosaurus,* probably still preferred the higher trees of its time. *Alamosaurus* is only found in the Javelina Formation in Texas, and there is a huge, broadleafed fossil tree in that formation called *Javeli-*

noxylon. By the end of the Cretaceous period, tall broadleafed trees had largely replaced the earlier dominant conifers. The Texas horned dinosaurs (ceratopsians) and duckbills (hadrosaurs) seemed to prefer the more open woodlands and plains of West Texas, where a preponderance of flowering plants awaited such dinosaurs and their heads-down method of grazing.

So, the dinosaur age saw several big changes in the available plant life. By the time of extinction that ended the dinosaur age, there was similar extinction of many dinosaur-age plant species. Life on earth is based on plants (herbaceous matter), and plant-eaters (herbivores), and plant-eater-eaters (primary carnivores), and plant-eater-eater-eaters (secondary carnivores). That is why paleobotany is an important part of the study of dinosaurs. Texas has several well-known paleobotanists, like Dr. Bonnie Jacobs of Southern Methodist University.

Dealing with Dinosaur Uncertainty

Finally, a word or two about how dinosaur paleontologists adjust their wording to deal with usually incomplete information. If a dinosaur can be confidently described from a plentitude of available fossils, a genus and species name (in Latin) can be given in a scientific paper describing the creature. An example would be *Tyrannosaurus rex*. If the available material is too insufficient to be certain as to species, then only a genus, for instance, *Tyrannosaurus* will be given. If the fossil evidence is so insufficient that some doubt exists, the creature may be referred to as "tyrannosaurid"or "tyrannosaurine." Even more doubt may introduce the term "morph," meaning "shape or form." An example might be "tyrannosauromorph"—what a word! Since many Texas dinosaurs are known from very little evidence, such qualifying terms are frequently used. In fact, at least half of the dinosaur paleontology of the future will be spent correcting honest, but mistaken ideas and identifications.

Most scientists try to do their best under the circumstances of a huge flood of new information. Some of the most important phrases among dinosaur paleontologists are: "Well, you got that right!" and "You almost got that right!" and most importantly, "We understand!" Texas dinosaur paleontologists do the best they can with what little they have. It gives future paleontologists a lot of opportunity and unanswered questions, and that is exciting.

Dinosaurs and Geology

Geology Grows Up

Geology is the study of the earth. When dino–saurs were first discovered, the science of geology was in a primitive state of development, compared with today. Because of this, many incomplete and mistaken notions have developed regarding the dinosaurs and their world. Today, geology makes sweeping new discoveries and refines its viewpoint at least yearly. In this book, we'll try to put dinosaurs in the most accurate geological setting that the constant changes of a still developing science can provide. We know full well that new information will alter our interpretations in the future. This changing information presents an important challenge and eye-opener to those who wish to study the earth and dinosaurs. New knowledge and changing interpretations will always be a part of science. A flexible, open mind and a sense of humor (irony) is critical in such work.

Fossils Tell an Important Story

The very existence of fossils, the remains or traces of ancient life, has taught us much about the earth. Few people today doubt that fossils actually represent ancient life. That was not always the case. There are still some people who dispute a particular fossil's age, but almost gone is the day when fossils were looked upon as tricks of the devil, put here to deceive mankind about the past. Even modern critics of the fossil-age dating purported by scientists are in agreement that fossils are remains or traces of the past, however long ago one conceives that to be.

The Fossil Record

Scientists have studied the earth long enough to know pretty well what kind of fossils occur in what rock layers. This is called the *fossil record.* Using this knowledge, we can match up distant, unconnected rock layers by whether or not they contain similar fossils. This rock layer matching is called *stratigraphic correlation.* We correlate rock layers around the world and put each in its proper place, along with the part of the fossil record it contains.

Thus we build a theoretical stack of rocks (a *stratigraphic column)* just as such a stack would look if all the earth's rock layers occurred in their natural sequence, with the oldest on the bottom and progressively younger and younger layers on top of it. We call this "oldest on the bottom, younger on the top" idea *superposition.* That is to say, a younger layer of rock will normally be found to lie in a super (top) position on older layers previously laid down. This remains true unless great movements within the earth—mountain building, for example—upset this relationship after the rock has been deposited.

It is only possible to build this stack in our minds and in theory. Nowhere in the world has every moment of time left layers of rock in such a logical and complete sequence. Erosion, for instance, can cause layers in some places to be worn away or keep sediment from being deposited at all. We find a bit of our stack over here and a bit of our stack over there and by using correlation and superposition, we can envision the total theoretical stack in our minds. When we build this in our imaginations, it is as if all the rock we know about anywhere is at its proper place in our one stack.

A well-known Cretaceous rock exposure near Fluvanna, Texas, being examined by Wayne Seifert, formerly of the Dallas Museum of Natural History, and Dr. John Brand of Texas Tech University.

What do we find when we examine the fossils in our stack? First, we notice that the earliest rocks that formed on earth, the oldest rocks at the very bottom of our stack, contain no fossils at all. In fact, except for faint traces of very simple algae-like fossils, the lower half of our rock stack shows little evidence of life beyond such simple forms. While microscopic, microbial life appeared fairly early in the earth's history, it changed very little for 3.5 billion years. Therefore, we must look quite high up in our stack, at an age of roughly 500–600 million years ago, to see the point where a seeming explosion in the diversity of fossils occurs. From that point on, as we move upwards in our rock column, we find it contains an increasingly greater variety of more complex organisms. This sudden appearance of a diversity of life-forms, represented by fossils, is one of the great mysteries of earth history. It is often called the *Cambrian Explosion,* after the Cambrian period of geologic time in which it occurred. Some scientists believe earth's plant life had only by then produced enough oxygen for such complex organic development. Many scientists are searching for other explanations (Dott and Prothero, pp. 209–212).

We would not have recognized this phenomenon without our theoretical stack of rocks and a knowledge of the fossils at various levels within it. One of paleontology's greatest achievements has been deciphering and constantly refining the fossil record.

A Ten-Story Building

If we think of our worldly stack of rock as a ten-story building, with the first rock formed on earth as the building's foundation, then the Cambrian Explosion of fossil forms would not be seen until we reach the ninth floor! As we pass upwards from there in the rock stack, fossils of animals without backbones (invertebrate fossils) are joined at a point by fish-like fossils. Higher still, fish (vertebrates) are joined by land vertebrates such as amphibians, and then by reptiles. In the analogy of our ten-story building, *all of dinosaur history would take place on the tenth floor.* Dinosaurs pass from the scene into extinction, leaving only a few layers of rock in our stack (the roof beams of our ten-story building) to represent all of the 65 million years since the dinosaurs disappeared. *All of human history is only the dust on the roof of our fanciful ten-story building.*

What can we learn about dinosaurs from this? Primarily, we can see that dinosaurs are one of life's more modern developments in the long history of the earth. Dinosaurs, even though gone for 65 million years, are a fairly recent life experiment that ended not so long ago in geologic

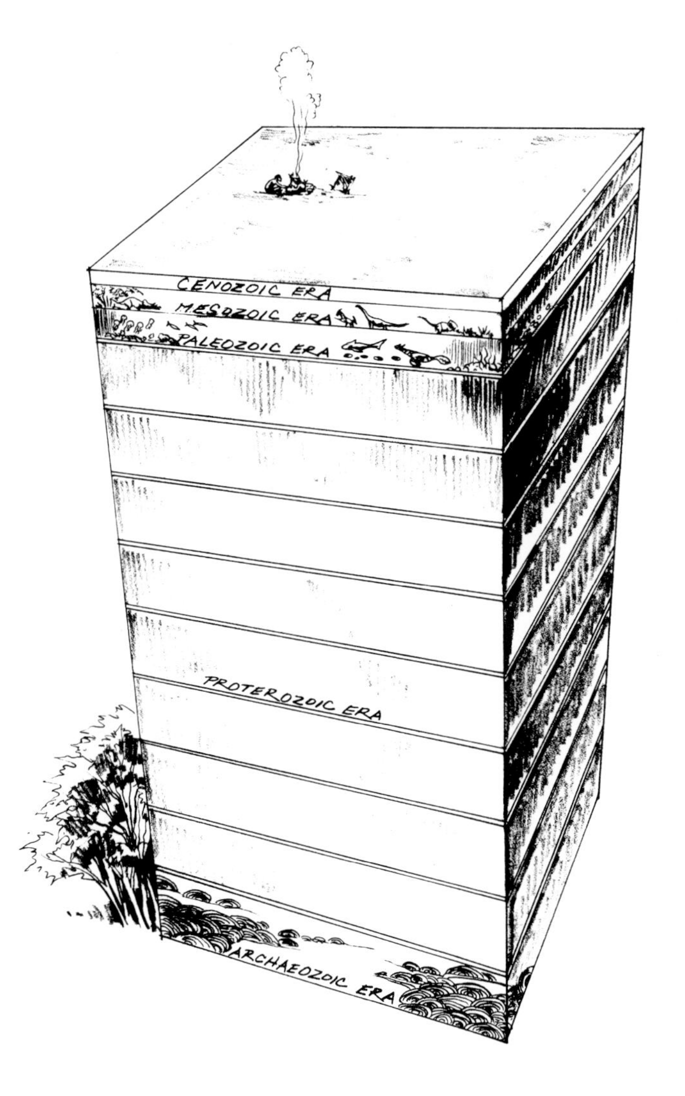

terms. Dinosaurs and our own human species belong to this recent end of the fossil record.

If it were not for our theoretical stack of earth's rocks, we could never know that dinosaurs first appeared at a certain time (the Late Triassic Period, about 230 million years ago), nor could we guess when they disappeared (the last of the Cretaceous period, about 65 million years ago). We could not say that fish and amphibians were on the earth long before reptiles. Nor could we say that dinosaurs and humans lived at separate times, if we did not see that dinosaur and human fossils never occur together in the same layers, and that human fossils appear only in layers laid down long, long after the layers containing dinosaur fossils. We have gained a great many insights from the fossil record and our theoretical stack of rocks.

If we think of earth history as a ten-story building, all of dinosaur history would take place on the tenth floor.

How Old Are Things?

Our stack of rocks, however, can only suggest the ages of things in a relative way, in relation to whether something occurs higher up (younger) or lower down (older) than something else. It cannot tell us just how much younger or older one thing is than the other. What we are doing with our stack of rocks is called *relative dating.* If we want to know within a specific number of years how many years apart two fossils lived, then we must find a "clock" in the earth. Such clocks have been discovered in the way some natural elements in the rock change in atomic composition regularly through time—like clockwork. The scientific search for such "clocks" is one of the most active and exciting activities of modern science. Dozens of natural elements have been found that do this. Most of them are *radioactive elements* and undergo a natural process of radioactive decay. That means they naturally give off atomic parts of themselves and change at a regular rate from a geologically formed unstable atomic structure into a more lasting, stable form. The starting point, the time of crystalization, implies that this procedure is most useful with crystalline rocks, which are rocks that were once molten (igneous rocks) or physically recrystallized (metamorphic rocks). A few sedimentary rocks are dateable, but with less precision.

If you know when such unstable, crystalline elements were first formed and can measure how much they have changed, you have a good clock. For dating fossils, however, it must be a slow-running, long-term clock, able to show changes of hundreds of thousands or millions of years. The well-known and much publicized Carbon 14 isotopic clock exhausts its useful changes in 80,000 years or less. It is more of an archeological (human prehistory) tool than a tool for dinosaur paleontology. The last actors in our dinosaur drama died 65 million years ago. The change, or decay, of unstable forms of uranium and potassium takes millions and billions of years, and is more useful for the long haul of time.

The various unstable and stable forms of these elements are called *isotopes.* The dates given in this book are *isotopic dates,* based on such measurements. Such dates are sometimes said to be absolute but, nonetheless, they are not perfect. All such dates are assumed to have a certain plus or minus factor, to allow for impure samples and testing error. Growing experience, better equipment, and cross-checking the results of one kind of "clock" against another make such dates ever more reliable. Enough different methods have been found to give independent dates on a particular sample that, if cross-checking is used, the results are less subject to criticism.

Drifting and Dreaming

Recent study of the earth teaches us another very important lesson. The earth's very skin (crust) changes through time. Much of the interior of the earth is a very thick, plastic, almost fluid, molten rock. We live on a thin, cooled shell of rock, a few miles thick, that we call the earth's crust. We recognize distinct parts, or plates, of the crust that form either continents or seafloor. The continental plates are made of lighter-weight rock material and move slowly over the molten material beneath them. Molten material wells up out of great cracks on the ocean floor to form a spreading, rearranging movement of the oceanic plates between the continents. In this process, the ocean plates either push the continents, collide with them, or often dive under them. The diving process is called *subduction.* The diving crust is reheated to a molten state in the depths of the earth. As an ocean plate dives to its fiery end, it may drag continental plates behind it. This welling up of molten material, the spreading of the ocean bottom, together with the forcing down of the leading edge of the ocean-bottom plates to be remelted is all a rocky treadmill. Its motion recycles the ocean bottoms. The continents are, in turn, constantly rearranged in patterns of togetherness and separation. One of the clues to this revolving pattern came as a result of scientists searching for very old sea-bottom deposits. It seemed to make sense that the ocean bottom would contain all of the sediment ever washed into it. To our great surprise, we found that the oldest rock in the depths was only about as old as the dinosaurs. The earlier rock had all been subducted into the earth, remelted, and was gone (unless some of it reappeared through volcanoes). Chains of volcanoes rise through fissures near the remelting zones, and mountain masses arise by slow but gigantic force where continents collide.

This broken surface of "plates" (ocean plates and continental plates) moving over the earth's molten interior and propelled by changes in the hot seething material below is constantly changing the very nature of our planet. We call this *plate tectonics* (plate movements), and the result we call *continental drift.*

This is a somewhat recent concept. I took a geology course in the early 1960s. The 1959 textbook for the course rejected all the evidences for continental drift by stating, "Though the theory is a brilliant *tour-de-force,* its support does not seem substantial." In less than five years after that publication, drifting continents were the most scientifically accepted explanation for this aspect of earth history. Continental drift was proposed, without much effect, in the early 1900s by American geologist F. W. Taylor and by German meteorologist Alfred Wegener. It grew in acceptance during and after World War II because of the confirmation of seafloor movement by wartime oceanic research. There is much to ponder in how great ideas like this are at first rejected by conventional wisdom, and then later come to acceptance. A dramatic idea such as floating continents forces us to ask what this meant to the world of the dinosaurs.

The dinosaurs witnessed some of the most dramatic of these drifting continental changes. Dinosaur-like reptiles evolved just as the earth was ending a period that saw only one great continent, surrounded by one great ocean. Geologists have given that supercontinent the name *Pangaea,* which means "all land." It was on that one land mass that pre-dinosaurian reptiles evolved. The first dinosaurs evolved as that huge land mass of Pangaea began to break apart. The newly separated continents started to drift toward their present-day positions. Dinosaur history, therefore, coincided with the shaping of the world we know today. Dinosaurs are creatures that rode the passing of an ancient geological world into a world of several oceans and widely separated continents. Although many kinds of dinosaurs evolved and became extinct throughout this process, there is nothing to suggest that the last of the dinosaurs had not adapted well to this new world and could not find a place or a way to live in it. As far as drifting continents are concerned, many kinds of dinosaurs seem to have coped with the changes. They were even stimulated by such challenges to evolve new species as their old groupings became separated by new and widening oceans, deserts, and mountain ranges. This new world was for a time *their* world. The mysterious extinction that loomed in their future is not necessarily linked to continental drift or plate tectonics. The important concept is that their world was a changing world, and some dinosaurs were changing with it as successfully as were our mammalian ancestors. The geographic changes were extremely gradual, taking place over a period of millions of years.

What About Texas?

What can we learn from this about Texas? For one thing, we can't expect the Texas of the distant past to be exactly like the Texas of today. Despite the fact that the rock under our feet seems cold, hard, and durable, we live, 65 million years after the dinosaurs, on a young, hot, changeable planet. Texas began the Age of Dinosaurs fairly near the earth's equator. As continental drift progressed, Texas moved further and further northward, as part of the North American continent. Hence, a more varied climate came about as time went by. In addition, Texas' connection with Africa, Europe, and South America (all of which were near us as fellow parts of Pangaea) were at times severed. Our connection with Asia was longest retained, and South America eventually rejoined us toward the very end of dinosaur times. So, we have long-standing relationships with Asia, while our newest influences have "just arrived" (geologically) from South America.

(text continued on page 68)

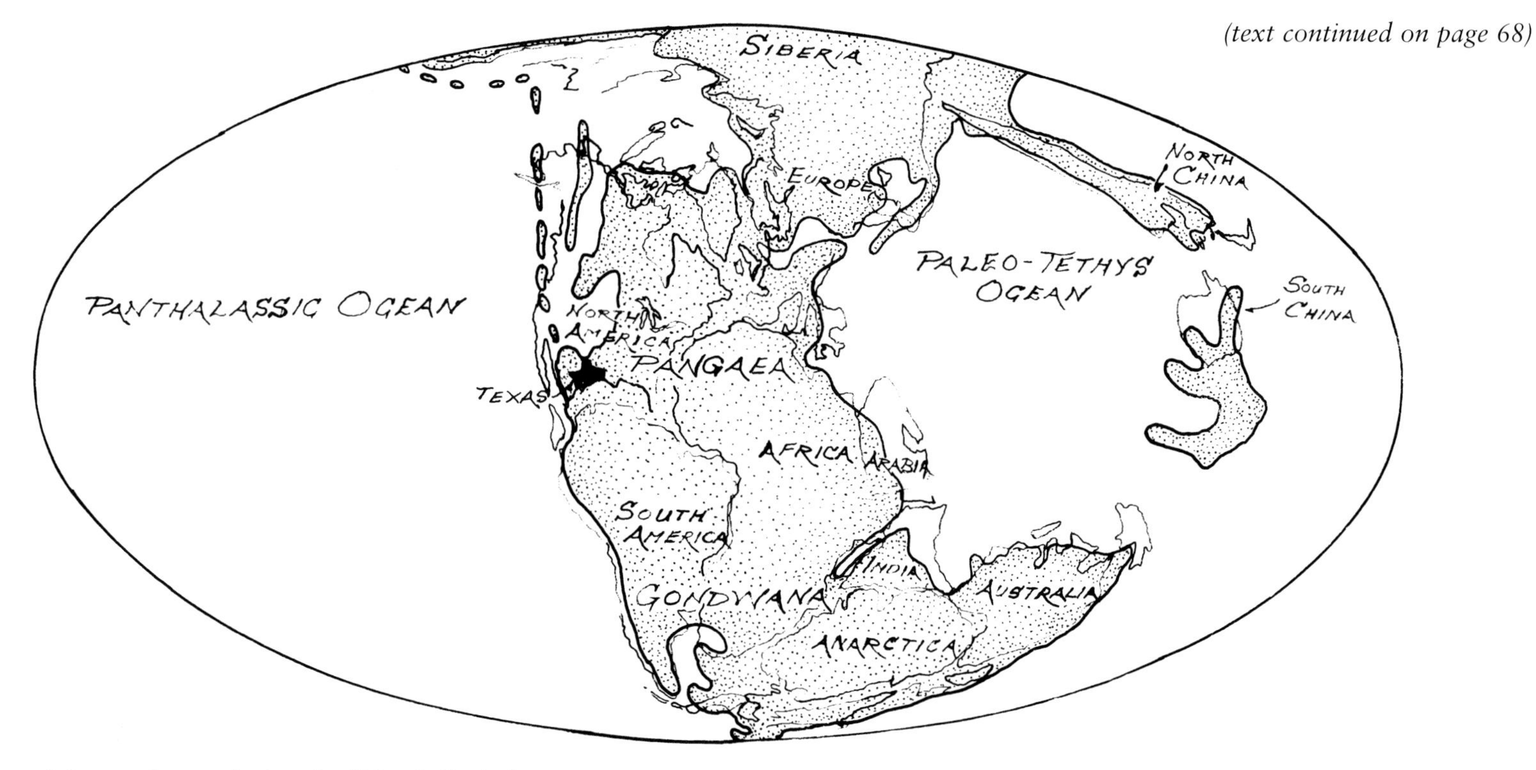

Position of the continents during the Triassic Period.

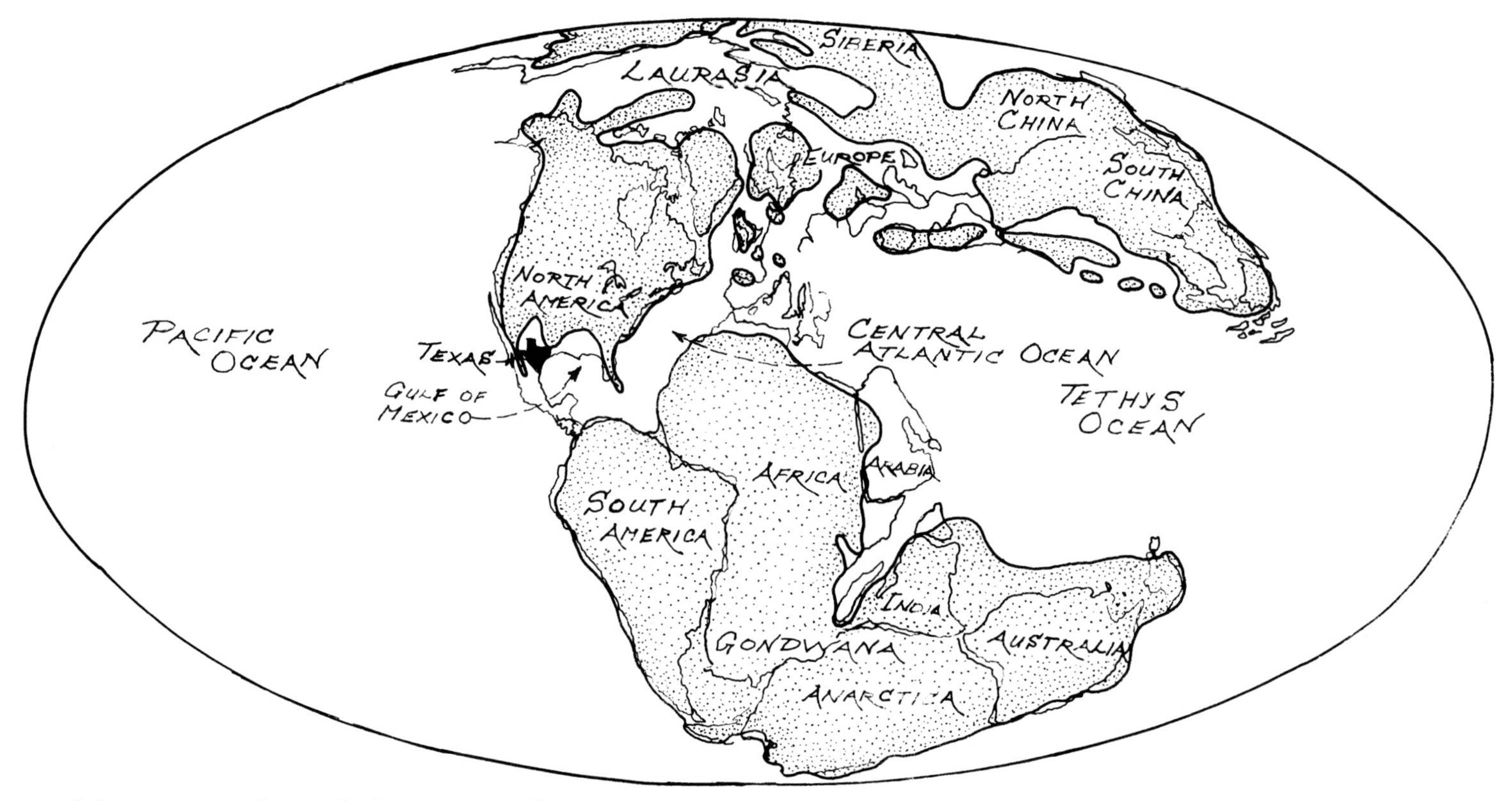

Position of the continents during the Jurassic Period.

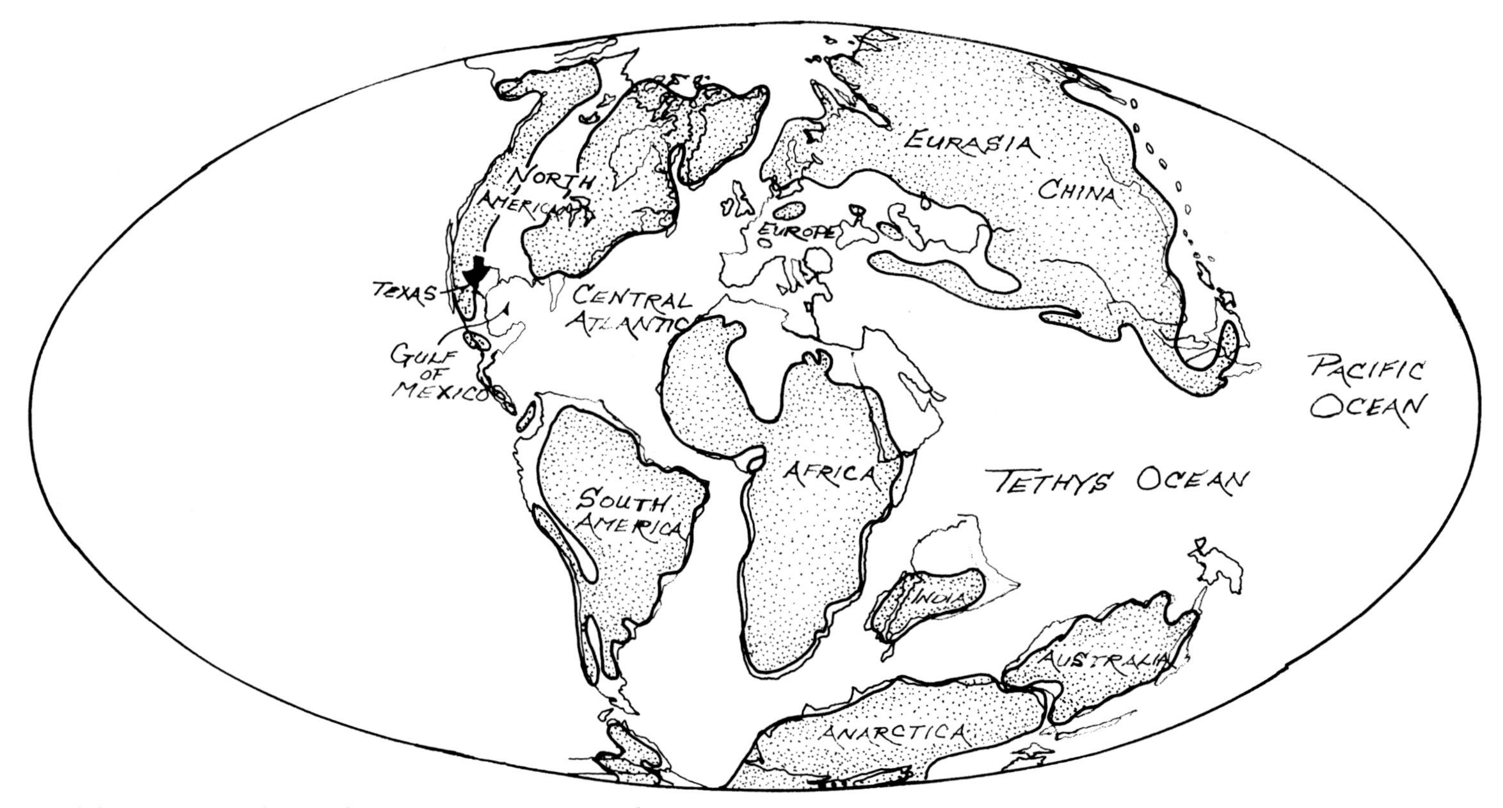

Position of the continents during the Late Cretaceous Period.

(text continued from page 65)

Dinosaur-Age Rock in Texas

Paleontologists call rock that surrounds fossils in the ground *matrix*. It has been the fossils' bed and protection for millions of years. Do we have much rock from dinosaur times in Texas? If so, where is it and what is it like?

Those very large chunks of geologic time we spoke of earlier as eras are made up of smaller units of time called geologic periods. Science recognizes three great periods of geologic time in the Age of Dinosaurs (the Mesozoic Era), from oldest to youngest: the Triassic, Jurassic, and Cretaceous periods. When a certain rock layer appears on the surface, where we can touch it or walk on it, we call that an outcrop or exposure. The rock is exposed to the open air instead of being buried under other rock layers. Geologists and paleontologists spend much of their field trip time searching for rock exposures and outcrops.

In an overall view, Texas has:

- many exposures of suitable rock bearing vertebrate fossils from the Late Triassic period
- very few rock exposures from the Jurassic period

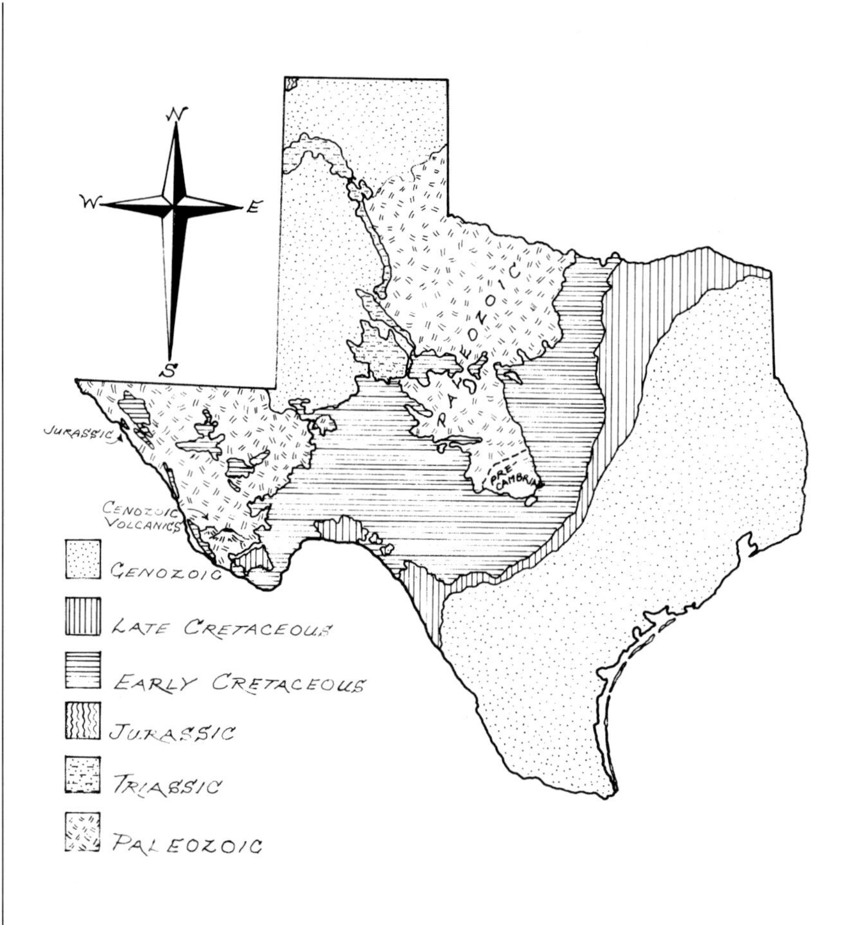

Rock outcroppings in Texas.

- wide exposures of marine (sea bottom) rock from the Cretaceous period
- considerable exposures of terrestrial (land-deposited) Cretaceous rocks formed in the right environments to contain dinosaur fossils.

Texas Triassic Rock

Exposures of reddish Triassic sandstone and shale occur as a north-to-south elongate exposure through the center of the Texas Panhandle, from the Canadian River to the Pecos River, virtually from Oklahoma to Mexico. In Texas, only rocks from the last part of the Triassic period (called the Dockum Group) are exposed, but fortunately for us, that is about when the dinosaurs first evolved. For studying dinosaurs, Texas' Triassic rocks have the potential of producing very interesting, very early dinosaurs. A large assortment of non-dinosaurian reptiles has also been collected from these beds, especially by paleontologists at the University of Texas at Austin, the University of Michigan, West Texas State University (now West Texas A&M, at Canyon), and recently by Dr. Sankar Chatterjee of Texas Tech University in Lubbock. All these institutions have exciting exhibits of the non-dinosaurian Triassic reptiles and the associated amphibians. The Late Triassic had a wide variety of reptilian forms, both dinosaur and non-dinosaurian. This is what attracts paleontologists like Dr. Chatterjee and others to study it. Dinosaurs were not so noticeable among other reptiles of that time, and certainly were not yet dominant. That is a time in which they had to compete with many other reptiles for supremacy. After the Triassic, dinosaurs fairly dominated the scene in the Jurassic and Cretaceous periods. After the Triassic period, it would have taken a major act of nature, such as the possible collision with a giant asteroid or a sweeping climatic change or disease (rather than just the competition with other groups) to bring the dinosaurs to any degree of total extinction.

Fossil discoveries, definitely identified as the bones of genuine dinosaurs, have proven elusive in the Texas Triassic beds. In New Mexico, complete skeletons of the little dinosaur, *Coelophysis,* are found in similarly aged rocks. It is not unthinkable to expect *Coelophysis* in Texas. Yet, the only remains ascribed to a Texas *Coelophysis* even tentatively are some isolated teeth.

A polyester replica of a New Mexico *Coelophysis* generally fills in for a genuine Texas *Coelophysis* in Texas museums, as is the case in the Panhandle-Plains Historical Museum in Canyon, Texas. Such casts do their educational job, but we would all like to find a real Texas *Coelophysis.* It is one of the reasonable expectations of Texas

Coelophysis *exhibit (fiberglass replica), Panhandle Plains Historical Museum, Canyon, Texas. This small Triassic meat-eater has not yet been identified in Texas. It is more commonly found in New Mexico.*

dinosaur paleontology. Dr. Phillip Murry, a Triassic specialist, feels we need a better understanding of Texas Triassic rock layers. He feels suitable layers for Coelophysis could exist here.

Preservation of anything approaching complete skeletons is rare in Texas Triassic rock. Its sand and claystone exposures are the result of freshwater stream deposition and often show wildly bedded (crossbedded) sand layers indicative of swiftly shifting stream currents. Such turbulent stream action tends to separate skeletal pieces. This has caused a lot of, "What belongs with which?" problems for Texas Triassic paleontologists. "Does that really go with that?" is a common Texas Triassic dinosaur question.

The Triassic period found Texas still embedded in the midst of the several continents, making up the parting days of the supercontinent of Pangaea. Most of Texas was dry land, above ocean level during Triassic times. Eroding highlands existed in East Texas. These were leftover uplands from the great mountains thrust up during the continental collisions that formed the supercontinent millions of years before. Can you imagine that Dallas, Waco, Austin, and San Antonio were all part of a towering mountain chain in late paleozoic times? Twisted remnants of those mountains still exist beneath the surface of Texas and above the surface in the eastern North America as the Appalachian Mountains.

The "state" sloped westward toward a wide basin area in what is today the Texas Panhandle. Texas streams flowed westward into that depression. Earlier, during the preceding Permian period, similar streams had helped fill in the Permian Basin in West Texas. Fossil deposits often occur in the channels of these westward flowing ancient freshwater streams. Freshwater mussels and fossil plant remains are plentiful in these meandering Triassic stream deposits.

Texas Jurassic Rock

The early part of the Jurassic period mostly continued the trend of the Triassic, with Texas being often high and dry with little opportunity for sedimentation and fossil deposits to form. The Jurassic, well-known elsewhere for famous dinosaurs like *Stegosaurus, Apatosaurus,* and *Diplodocus,* is feebly represented by marine rocks in the Malone Mountains of far western Texas. Even those, Dr. Keith Young of the University of Texas at Austin tells me, are likely to be geologically thrust across the "border" from Mexico. In any case, they contain no remains of dinosaurs. Other fossils do indicate that some Malone

Mountains rock, at least, was formed by a shallow late Jurassic sea.

A recent contention in a UT Dallas masters thesis (Tickner, 1987) claiming that the Malone Mountains are completely Cretaceous is based, in my opinion, on insufficient evidence of a very few stray microfossils. I am supported in that opinion by Dr. Keith Young, who has studied the Malone Mountains for years. I must admit, I am interested in preserving the reputation of what little Jurassic rock we have in Texas.

A tiny exposure of terrestrial (land-deposited) Jurassic rock is reported by geologists in the extreme northwestern corner of the Texas Panhandle, but no well-documented dinosaur bones have yet been found there. Rumors abound among paleontologists that such exposures do exist, but the extent of such exposures is certainly small and doubtfully productive. That is an area of the Texas Panhandle where land access for collecting has been difficult to obtain. It has very large ranches and many absentee landowners. Note: Always seek *permission* before entering private land anywhere, *especially* in the wilds of the Texas Panhandle.

Jurassic dinosaur fossils have been found in abundance not far from Texas, in Oklahoma and New Mexico. We may

Dr. Keith Young, UT Austin expert on ammonite zones in Texas rock, with an ammonite from the Malone Mountains, Hudspeth County, Texas.

(text continued on page 76)

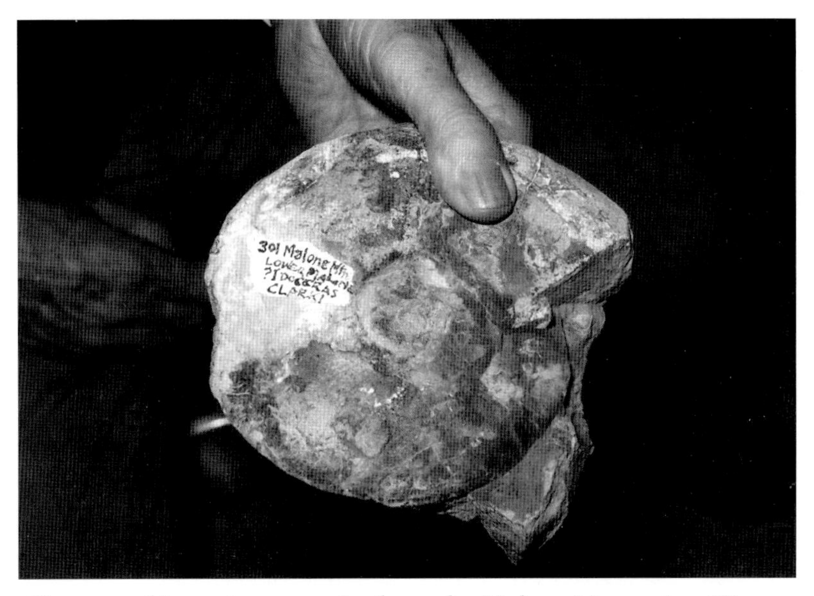

Close up of Jurassic ammonite from the Malone Mountains, West Texas.

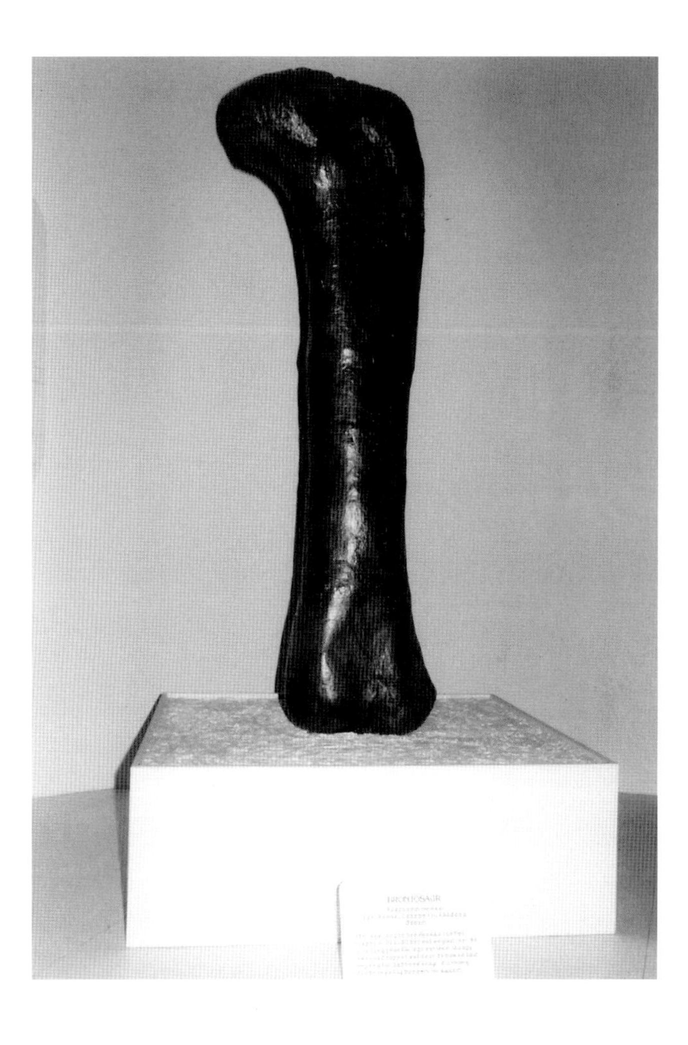

Brontosaur, Apatosaurus, *leg bone (femur), from the Oklahoma panhandle, displayed in the Panhandle-Plains Historical Museum exhibit, Canyon, Texas. It is the only Texas exhibit of Jurassic material from Oklahoma.*

This 70-foot Jurassic Diplodocus, *at the Houston Museum of Natural Science, is not a native Texan, but is the state's largest assembled dinosaur. With it are shown casts of two Cretaceous period dinosaurs,* Tyrannosaurus *and the duck-billed dinosaur* Edmontosaurus. *Both of these have close fossil relatives in Texas.*

A Jurassic Allosaurus *(cast) dominates the fossil gallery of the Panhandle-Plains Historical Museum, Canyon, Texas.* Allosaurus *is not native to Texas.*

A most aesthetic dinosaur exhibit, this Jurassic Allosaurus *and* Camptosaurus, *Fort Worth Museum of Science and History, is one of the oldest such exhibits in the state. Neither species is native to Texas.*

Jurassic dinosaurs (Allosaurus, Diplodocus, Apatosaurus, *and* Stegosaurus) *are native to the Oklahoma Panhandle, but are only suspected in the northwest Texas Panhandle.* ▼

(text continued from page 72)

hope, therefore, that Jurassic dinosaur bones will yet be discovered in the northwestern corner of the Texas Panhandle. Meanwhile, we can only envy our neighbors to the north and west for their wealth of Jurassic dinosaur remains.

Several Jurassic dinosaurs from other states grace the halls of Texas museums. Since they are adopted Texans, we shall include photographs of some of them. In fact, the largest dinosaur on exhibit in Texas, the giant sauropod *Diplodocus* at the Houston Museum of Natural Science is a non-Texan Jurassic dinosaur.

Texas Cretaceous Rock

The most recent and last of the three periods of the Mesozoic Era, the Cretaceous period, is the most widely exposed across Texas. Cretaceous period rock exposures in Texas are very plentiful, but the majority of them are marine limestones and shales deposited under shallow seas. During the Cretaceous period, overflow waters from the ancient Gulf of Mexico shallowly submerged parts of Texas. Sea levels rose around the world, and shallow Cretaceous seas spread inland across central North America, Asia, and Africa. This has been linked to the greater output of molten rock in mid-ocean ridges, as more active continental breakup and drift followed the almost static plate movements of the mass of continental plates making up the supercontinent Pangaea. It is thought that the insulating effect of the one large land mass caused an accumulation of heat energy beneath it. This was then followed by extra active welling up of molten rock.

One of the major breaks in Pangaea was the opening of the North Atlantic Ocean, with greatly increased activity along the mid-Atlantic ridge. With more lava in the ocean depths, there was naturally a displacement of sea water. The results are clear. Sea levels rose and fell and rose again many times during the Cretaceous period. Texas went from mostly dry to just about entirely sea-covered and back to fairly dry again, before the end of the dinosaur age. At their highest and longest rise, seas covered the low central areas of North America, splitting the North American continent into an eastern and a western part with the shallow sea in between about where the Rocky Mountains stand today.

This marine environment preserved many invertebrate fossils, some of the finest and most plentiful in the world; but it was an inappropriate paleoenvironment for finding the remains of dinosaurs. Fortunately, there are other Texas Cretaceous deposits of sandstones and shales that

(text continued on page 80)

Dinosaurs were big, but this ammonite Parapuzosia bosei *at Texas Christian University is a yard across. Four-foot specimens are known from Texas marine Cretaceous rocks. Ammonites became finally extinct at much the same time as non-avian dinosaurs.*

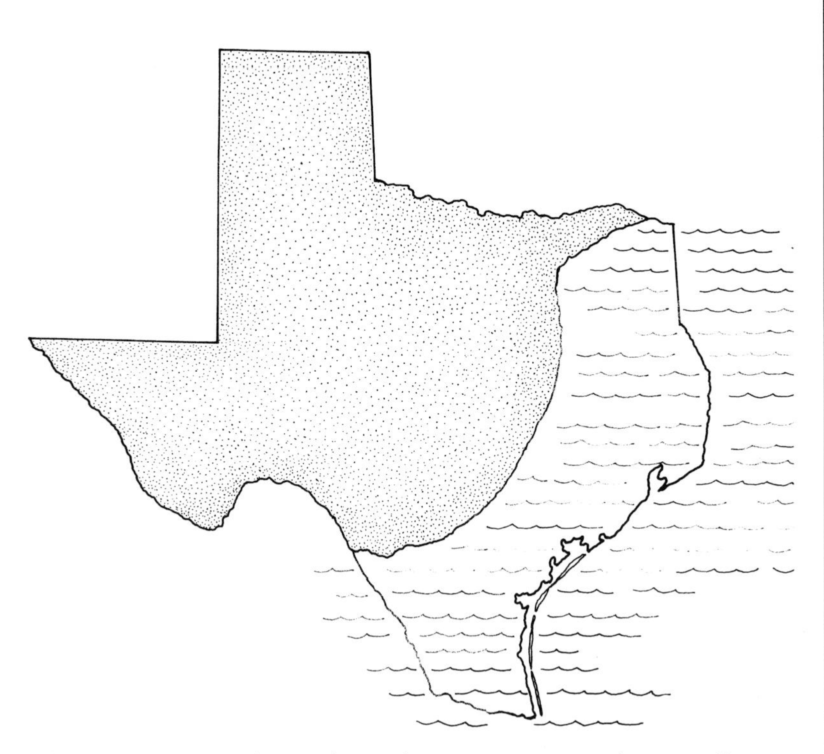

The Cretaceous sea during the Early Cretaceous Period, 115 million years ago, the time of the Twin Mountains Formation.

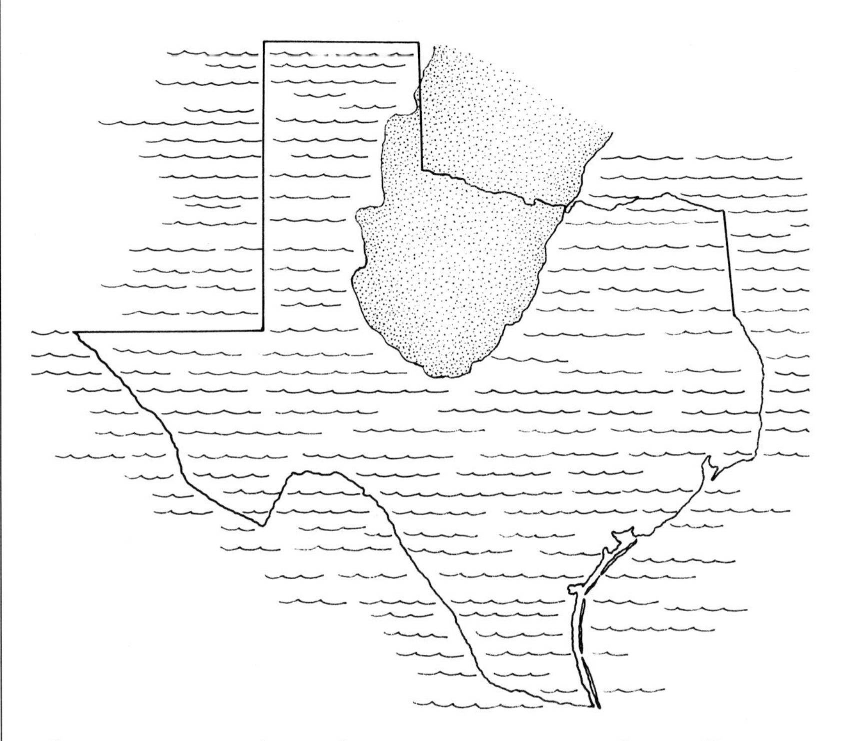

The Cretaceous sea during the Late Cretaceous Period, 80 million years ago, the time of the Austin Formation.

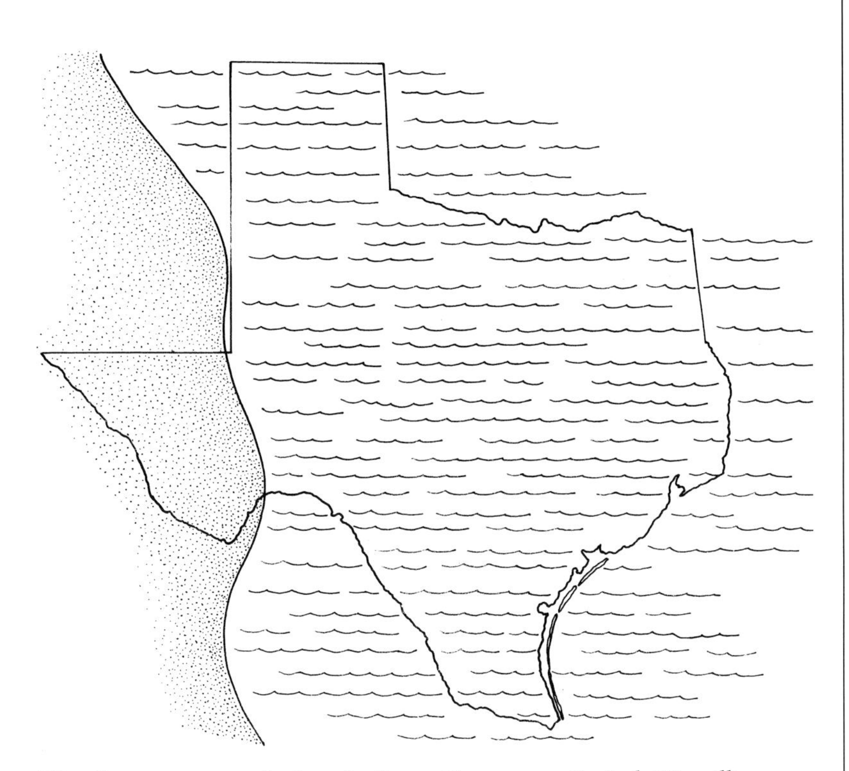

The Cretaceous sea during the Later Cretaceous Period, 70 million years ago, the time of the Aguja Formation.

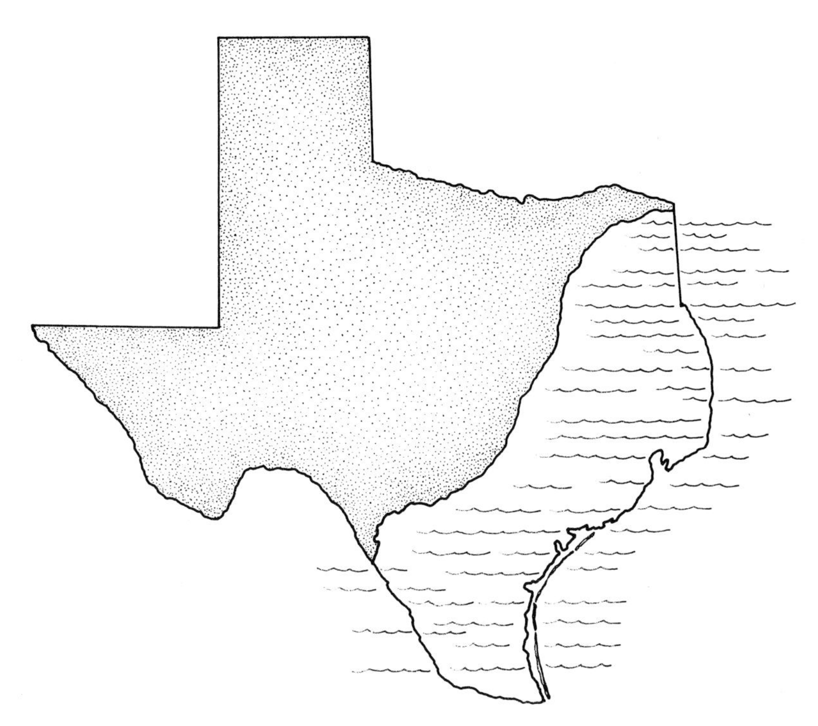

The Cretaceous sea during the very Late Cretaceous Period, 65 million years ago, the time of the Javelina and Nacatoch Formations.

(text continued from page 76)

represent a dry land or shoreline environment. It is in these that dinosaur remains are found.

Unlike the Texas landscape of earlier times, by the Cretaceous period, Texas topography (the high and low elevations) had become more like today. The rivers were flowing southeastward, as they do today, not westward as they did during the Triassic period. The seashore was restricted to East Texas during the first half of the Cretaceous period, sometimes moving closer to central Texas and sometimes retreating southeast.

North Central Texas

In north central Texas (and into Oklahoma and Arkansas), some of our oldest Cretaceous rock is exposed in a thick series of sandstone layers, deposited some distance from the sea. In northern Texas and Oklahoma, it is called the Antlers Formation. In Arkansas, it bears the older, general name of Trinity Sands. Further south, in north central and central Texas, this same formation is split into upper and lower sections by a mostly limey deposit, the Glen Rose Formation. The Glen Rose represents a spreading inland over central Texas of a shallow edge of the sea for a geologically brief period (a few million years) during mid-Antlers times. So, in central Texas, the part of the Antlers Formation below the Glen Rose limestone is separately named the Twin Mountains Formation. The sandstone above the Glen Rose incursion is called the Paluxy Formation. The marker created by the presence of the Glen Rose limestone means that fossil finds from these layers in north central Texas can be given a relative age with more precision than further to the north. In north Texas, the Antlers Formation (Twin Mountains, plus the Paluxy Formations) looks pretty much the same from top to bottom.

The Glen Rose limestone, formed during that middle period (in north central and central Texas), when the laying down of the Twin Mountains sand was interrupted by an incursion of the sea, provided a unique environment of coastal lagoons and tidal flats. Many dinosaurs *(Pleurocoelus* and *Acrocanthosaurus,* in particular) left their footprints in the limey muds of the Glen Rose Formation tidal flats, which extended for great distances. These older Texas Cretaceous sediments: Antlers, Twin Mountains, Paluxy, or Glen Rose, all contain discoveries of dinosaur fossil bone. These formations are approximately 120–115 million years old. It should be noted that in central Texas, rock of the

Brontopodus birdi footprint, probably made by the brachiosaurid sauropod Pleurocoelus, on exhibit, Dallas Museum of Natural History. Track was excavated by R. T. Bird at Glen Rose, Texas, 1940. Viewing the track are Mr. Avelino Segura and Brian Barnett of the DMNH staff.

Texas dinosaur footprint from a three-toed meat-eater, perhaps Acrocanthosaurus. Dallas Museum of Natural History.

same age as the Twin Mountain Formation is locally called the Travis Peak Formation. The Travis Peak Formation has not provided much in the way of dinosaur fossils.

Worldwide, the Cretaceous period began about 135 million years ago. That means Texas' oldest Cretaceous rocks omit any record of the first 15–20 million years of Cretaceous time. Commonly, our earliest Texas Cretaceous is more equivalent to the mid-Cretaceous on a global scale. If I refer to it ever as Early Cretaceous, I mean in Texas.

The Intriguing Bissett Formation

One different perspective on the Texas Cretaceous is provided by the recent change in thinking about the age of the Bissett Formation, which outcrops along the north and west sides of the Glass Mountains in Brewster and Pecos counties of West Texas. The Bissett was originally thought to be Permian Period in age. Then, for many years it was thought to be Triassic, based largely on only photographic evidence of a fossil thought to be a part of the non-dinosaurian reptile *Desmatosuchus,* which would indicate the Triassic period. Field research in 1989 by Robert Wilcox and Dr. David Rohr of Sul Ross University in Alpine, Texas, discovered vertebrae of the Early Cretaceous ornithopod dinosaur, *Iguanodon,* in rocks of the Bissett Formation. Dr. Wann Langston, Jr. made the identification. The specimens are housed in Austin. These vertebrae are very similar to vertebrae of *Iguanodon* from England. Such creatures usually indicate, in the European

Iguanodon *vertebrae, Bissett Formation, Brewster and Pecos counties, Texas. Prepared by acid bath and careful chipping at UT Austin, these fossils changed opinion of the age of the Bissett Formation from Triassic to Cretaceous.*

system, the early Aptian Stage of the Cretaceous. Invertebrate fossils in the Bissett confirm the same. A lower member (layer) of the Bissett may be as old as the European Neocomian stage. This would make the oldest Bissett Formation Texas' oldest Cretaceous surface rock formation, at nearly 125 million years old. It also further pegs down *Iguanodon* as a Texas dinosaur species. But isn't it interesting how a whole geologic formation can be mistaken, reevaluated and then reevaluated yet again, as more accurate evidence comes to light?

The All-Across-Texas Sea

Following the deposit of these older Texas Cretaceous formations, the sea slowly made a major advance across Texas, eventually covering the entire state during most of the last half of the Cretaceous Period. The sea covering Texas extended all the way, south to north, across the center of North America. Dinosaurs were virtually unable to find a dry place to live in Texas for most of the last half of the Cretaceous period. Much of Texas remained beneath the surface of this shallow sea for 45 million years. Sometimes, I have the impression that people think the Cretaceous sea just came and went overnight. It was a major event in Texas prehistory. It *was* Texas (as in,"all there was") for tens of millions of years. No wonder, Texas is such a great place to find fossil mosasaurs and plesiosaurs, the aquatic reptiles of the time.

There were, however, at least two exceptions to this watery business. The first exception occurred when the eastern edge of the "all-across-Texas sea" retreated to expose some of northeast Texas, about 100–95 million years ago. That provided a temporary shoreline, just north of the Dallas-Ft. Worth metroplex.

Exposures of rock from this mini-retreat of the sea contain dinosaur tracks and bones. Such fossils are found in the Woodbine Formation and the slightly older beds of the Pawpaw Formation. This little, 5-million-year (give or take) temporary retreat of the sea seems to have occurred nowhere else in Texas. The resulting environment was lagoons and coastal tidal flats out of the sea for that geologically brief time. The dinosaur species found there have been mostly hadrosaurs (duck-billed dinosaurs) and armored nodosaurs (armadillo-shaped dinosaurs). Teeth of several unknown meat-eaters were also found there. Then about 95 million years ago these North Texas dinosaurs, who were enjoying a brief respite from the sea, were again evicted from their shoreline by the rising sea level. It would be 30 million years before that area would be dry again.

Some geologists have attributed this respite from the Cretaceous sea in northeast Texas to volcanic activity, perhaps even a volcanic island that rose from the seabed. Such volcanic islands are known elsewhere in the Texas Cretaceous period.

The second case of retreating sea was in reality the final retreat of the sea, toward the end of the Cretaceous period (about 75–65 million years ago) in both the western and the northern parts of Texas.

This last retreat of the sea produced the dry land needed for the very productive dinosaur fossil beds of the Big Bend region. The dinosaur bones in the Big Bend date from this Late Cretaceous time period, 75 million years ago, and lead right down in time to the extinction of all the dinosaurs, 65 million years ago.

Dinosaur Extinctions

Because we all wonder what finally happened to the dinosaurs, it is very special that Texas has fossils from those interesting days when dinosaurs unwittingly faced their end. It was the time of tyrannosaurs and duckbills and many-horned dinosaurs in greater numbers than ever. Even a giant sauropod, *Alamosaurus,* lumbered up from somewhere down around South America just in time to catch the end of the Cretaceous show.

Those who prefer the crash of a giant asteroid from outer space as a curtain closer for the Age of Dinosaurs find a tempting crater in Mexico (Hildebrand, 1991). It is just across the Gulf of Mexico from Texas. So the Texas dinosaurs were in front-row center seats for that shattering performance. All we can say for certain is that they were there and probably so was the giant crash.

The most recent thinking about the end of the dinosaurs tends to accept the fact that meteors, asteroids, and comets did, and do, collide with the earth. But paleontologists are less willing to accept that such collisions were the only problem dinosaurs faced in their final years. Climate, disease, and small extinction after small extinction all took their toll. The final dinosaur extinction is today viewed as a multiple-cause event, probably occuring over a few million years of time. Let us not forget, however, that as much as science is tending to take the awe out of the end of the non-avian dinosaurs, there was a last dinosaur and a last moment for their entire age on some unknown night 65 million years ago. It could have happened in Texas!

A Close Look at Big Bend

The Big Bend dinosaur beds contain a wide variety of species: *Tyrannosaurus* (a powerful carnivore), *Chasmosaurus* and *Torosaurus* (horned dinosaurs), *Alamosaurus* (the last big sauropod), as well as *Kritosaurus* and *Edmontosaurus* (the well-known duck-billed dinosaurs or hadrosaurs), and scraps of fossils of many species not yet well-identified.

In the early part of this final emergence of shoreline, the Big Bend was at the southern tip of that strip of western North America lying west of the mid-continental sea. This vast, shallow sea still separated eastern from western North America, but drier times were coming. The western part of North America was like a long, north to south, island. There was, more or less, continuous dry land all the way from the Texas Big Bend to Asia. The Big Bend region was at first home to a number of dinosaur species that wandered down this island from the north. The horned dinosaurs came to Texas this way. Fossils from this time are preserved in rocks of the Aguja (ah-GOO-ha) Formation. The rocks of the Aguja Formation contain evidence of sharks and rays and other fossils that indicate it

Tyrannosaurid tooth (TMM 41541-2) from Texas Big Bend, Brewster County, Javelina Formation. Probably not T. rex, *but a close relative. This tooth is about as long as your index finger.*

A horned dinosaur leg bone (femur), possibly Torosaurus *(less likely* Triceratops), *from the Texas Big Bend on loan from Texas Memorial Museum in Austin and displayed in the Witte Museum, San Antonio. Viewer is visitor Bryce Hubbell.*

and South America were still unconnected. It appears Texas was about as far south as you could go on dry land during Aguja times. During the later-deposited Javelina Formation there is evidence that a land bridge or chain of islands provided a means of interchanging dinosaur species between North and South America. *Alamosaurus*, the huge brontosaur-sized dinosaur, had its closest relatives in South America. No such dinosaurs existed in North America until Javelina times. It is natural to assume *Alamosaurus* emigrated into Texas from South America. It is found only in rock of the Javelina Formation, and not in the somewhat earlier Aguja Formation. The same is true of the horned dinosaur *Torosaurus* and a duckbill that looked much like the northern genus, *Edmontosaurus*. Likewise, the Aguja Formation is the only Texas rock layer to contain the horned dinosaur *Chasmosaurus* and the duck-billed dinosaur *Kritosaurus*. Thus it is clear that the two dinosaur fossil-bearing rock formations of the Big Bend are from different geological times, represent very different environments, and contain somewhat different species of dinosaurs. First, dinosaurs were able to reach the Big Bend only from the north during Aguja times, then later they could come as well from the south in Javelina times.

Overlooked Northeast Texas

Simultaneous with the later stages of this retreat of the sea in the Big Bend (Javelina times) was a matching retreat in the northeast corner of Texas, in the Nacatoch Formation. Let's examine northeast Texas during late Cretaceous times. As far as I know, this is the first dinosaur book to mention dinosaur possibilities in northeast Texas at the very end of the Mesozoic Era, a time very comparable to that of the Javelina Formation in the Big Bend.

For many years, people have been collecting fossils in the Nacatoch Sand, a formation in the Late Cretaceous, Navarro group. There are exposures of it in Kaufman County, not far east of Dallas. When one is standing in mudstone and sand of Mesozoic Age, the thought of possible dinosaur fossils comes naturally. Nonetheless, it has for years been considered that the Nacatoch was deposited too far out to sea to contain any real chance of dinosaur fossils. After all, some of it contains ammonites and other plentiful marine creatures.

Papers written by Dr. Tom Lehman, of Texas Tech University, and Mary K. McGowen and Cynthia M. Lopez confirm that the Nacatoch Formation contains not only marine deposits but also delta and tidal flat sediments. That opens up for all of us the possibility of a new dinosaur-friendly rock formation in northeast Texas. That would be a lot closer to many of us than the Big Bend.

Since the Nacatoch is better exposed where it extends into Arkansas, I checked with Dr. John Thurman, professor emeritus, at the University of Arkansas at Little Rock. Although dinosaur fossils of older Cretaceous rocks (Trinity Sands or Antlers Formation) have been found in Arkansas, dinosaurs in the Arkansas Nacatoch have no firm record.

Discoveries in the Nacatoch would be some distance from other dinosaur discoveries of that time period, and it is also close to the same time as *Tyrannosaurus*. Who knows what interesting Late Cretaceous dinosaurs may be awaiting a lucky collector in the Nacatoch Sands? It opens up some interesting future collecting possibilities. This is where amateurs, with more time on their hands, could be of service to the professional community. Just remember: find it, mark it, and contact someone who can properly handle such new information.

There were many vacillations of the Cretaceous sea, and the possibility always exists of "float" material from dinosaurs that died at the water's edge and drifted out to sea. They may have sunk on some unlikely marine sea bottom. I am sure this is not the last "new" possibility we shall see for future dinosaur collecting in Texas. The photo on page 51 shows an unlikely duck-billed dinosaur fossil from northeast Texas.

Eleven Chances to Find Texas Dinosaurs

So the story of rock that contains dinosaur fossils in Texas is a series of "snapshots" from perhaps eleven distinct instances in time. They are:

1. the Late Triassic Dockum group, mostly in the Texas Panhandle
2. a little bit of seldom-explored Jurassic Morrison Formation in the extreme northwestern Panhandle
3. the Cretaceous Twin Mountains Formation in northcentral Texas
4. the Cretaceous Glen Rose Formation in central Texas
5. the Cretaceous Paluxy Formation in north and central Texas

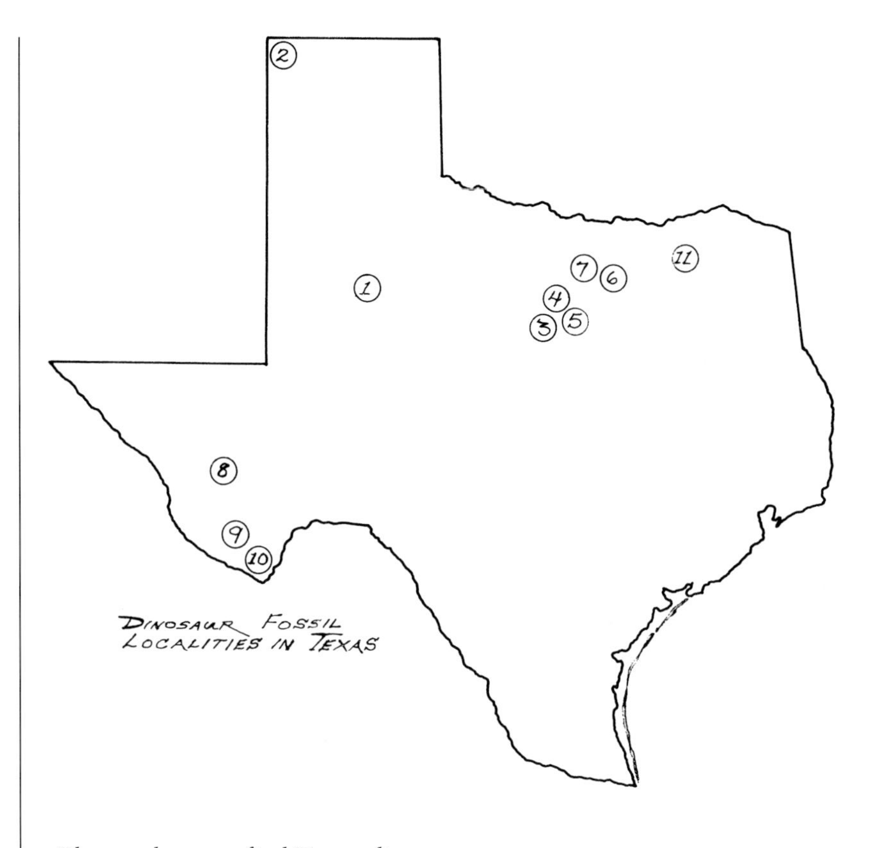

Eleven places to find Texas dinosaurs.

6. the Cretaceous Woodbine Formation in northeast Texas
7. the Cretaceous Pawpaw Formation in north Texas
8. the Cretaceous Bissett Formation in the Glass Mountains
9. the Late Cretaceous Aguja Formation in the Big Bend
10. the very Late Cretaceous Javelina Formation in the Big Bend
11. the very Late Cretaceous Nacatoch Formation, in northeast Texas

These provide plenty of places for future Texas dinosaur hunters to search. It is enough to allow a guess that eventually the total number of known Texas dinosaur species will double or triple. The rocks and fossils await you now, as they have for millions of years.

A geologic map, such as can be obtained through the Bureau of Economic Geology, University of Texas at Austin, Austin, TX 78713-8924, will lead you to such formations. Remember always to get land permission.

Making the Bones Bare

Who Was First?

Who found the first dinosaur fossil in Texas? A young man in the late 1870s, R. T. Hill, an amateur fossil collector in central Texas, counted in his collection two "saurian teeth." Hill took them off to college with him at Cornell University in New York State, where they were placed in the college collections. They have not been seen in recent years, despite much searching.

R. T. Hill was born in Tennessee and moved to Waco, Texas, as a young man to work in his brother's printing shop. He was a fossil enthusiast, like so many fossil collectors we have in Texas today. He was educated at Cornell University, where he learned the methodologies of professional geology and paleontology. He returned to Texas and in 1888 established the first curriculum in geology at the University of Texas in Austin. He quit the university after a couple of semesters because, among other indignities, the regents refused to buy him a proper geological microscope. He was apparently loved by his students and tolerated by many of his professorial colleagues. Hill was most at home in the field, collecting fossils and observing Texas geology firsthand and often alone.

R. T. Hill mapped and studied the rock layers and fossils of Texas to such an extent that he was appointed Texas state geologist. He founded the Texas Geological Survey, and has been called "The Father of Texas Geology." His own writings mention finding dinosaur bones near Millsap, in northcentral Texas, in 1876. Louis Jacobs, in *Lone Star Dinosaurs*, believes 1876 is a misprint and the actual date is 1886. Hill's fossils from Parker County were examined by the dinosaur man-of-the-day Edward Drinker Cope; but, like Hill's earlier "saurian" teeth, no trace of them has been found. R. T. Hill is the first collector to leave a record of finding dinosaur fossils in Texas. It is likely other amateur collectors made even earlier finds, but they did not maintain a lifelong career writing and studying geology as R. T. Hill did. This points out, all too clearly, the absolute importance of numbering specimens and keeping good records. There probably were many early Texas fossil collectors whose accomplishments are unrecognized today because they did not write down the details as well as did R. T. Hill. Make notes, keep records, take photographs.

A popular paleozoic pre-dinosaurian, Dimetrodon, *prepared by Kyle Davies, on exhibit at the Texas Memorial Museum, Austin, Texas. This specimen was collected during depression-era (WPA) work projects by Raymond Miller and Edgar Gardner for the Bureau of Economic Geology and the University of Texas.*

The Permian Distraction

The earliest interest in vertebrate fossils in Texas was in material even older than the dinosaurs, fossils of the Texas Permian "redbeds." Occurring on the surface in north Texas, these reddish-colored sands and claystones were discovered in the 1870s to contain plentiful vertebrate fossils, especially previously unknown early reptiles. From that time to the present, many collectors have sought such remains. Jacob Boll, W. F. Cummins, Charles Sternberg, E. C. Case, S. W. Williston, A. S. Romer, and others brought national attention to these productive pre-dinosaurian fossil beds. The great dinosaur hunter Edward Drinker Cope was most interested in Texas because of these Permian reptiles.

Indeed, Texas is still more widely known among the world's paleontologists for these more ancient fossil reptiles

The Dimetrodon *is often incorrectly called a dinosaur. Even though a well-known Texas fossil creature, it totally preceded the dinosaurs during Paleozoic times.*

of the Texas Permian than for its dinosaurs. Wherever one travels around the world, one can see on exhibit in the great museums the spectacular skeletons of these bizarre, early Texas reptiles, which are often incorrectly called "dinosaurs." One even sees the fin-backed *Dimetrodon* masquerading on dinosaur T-shirts. Nonetheless, it is nice to see that some other ancient Texas reptile besides a dinosaur can be famous.

Dinosaur Work in Big Bend

J. A. Udden, in his 1907 report on the Big Bend, lists identifications of several dinosaur bones by Dr. S. W. Williston of the University of Chicago. Williston's identifications and descriptions show clearly that these bones belonged to hadrosaurs, ceratopsians, sauropods, carnivorous dinosaurs, and the giant crocodilian *Deinosuchus.* It was, however, through the work of Ross Maxwell, studying the geology of the Texas Big Bend, that dinosaur bones in that area of West Texas finally attracted serious attention.

Among Maxwell's assistants was a graduate student, Hugh Eley, from the University of Oklahoma. Eley was interested in invertebrate fossils, but while helping Maxwell with his geological studies, he noticed large numbers of dinosaur bones in the Aguja Formation. Eley noted the presence of these bones on the geological maps Maxwell was preparing. In 1936, he called this to the attention of another OU student, William S. Strain. Strain's professor, J. Willis Stoval, was instituting a program of vertebrate paleontology at the university. Strain graduated and moved on to the Texas College of Mines and Metallurgy (now UT, El Paso). He taught geology and paleontology there for the next four decades. Having a job in West Texas, Strain immediately went to work investigating Eley's reports of dinosaur bones. He quickly discovered that Eley had understated the quantity of such fossils, which in some places literally covered the ground.

As it happened, that was during the Great Depression of the 1930s. The economic Great Depression caused the creation of work programs, which were sometimes put to good use excavating fossils. In this way, the "hard times" were actually a good thing for paleontology. The workers were accustomed to hard work, and excavating fossils was much more interesting to some than digging ditches. That period of government work removed so many fossils from the ground that fossil laboratories have not yet caught up with all the bones and teeth collected over half a century ago. Some of which, for lack of funds, still sit unexamined in several Texas laboratories.

The important thing is that the fossils were saved from certain destruction by weathering, and will now be available for study in future years, however long that may take. Fossil collecting by museums is a form of intervention in the deterioration process and conservation in its truest sense, even if the conservator may never live to see the final analysis of his work.

High in the Chisos Mountains, Ross Maxwell's helpers in the Big Bend State Park started a small museum at their headquarters in the Basin, which is the center of the Big Bend National Park today. O'Riely Sandoz, another OU student working with Ross Maxwell, was instrumental in setting up this small fossil exhibit in one of the wooden work buildings. Unfortunately, a fire destroyed the wooden building on Christmas Eve 1941. Most regrettably, all of the burned material, including many charred fossil bones, went into the junk heap. Little effort was made to salvage any of it.

In the 1930s, the University of Oklahoma began to work with the Works Progress Administration (WPA), another depression era jobs program, to make large fossil collections in Oklahoma. This again caught Strain's attention at the Texas College of Mines in El Paso. He obtained assistance from the WPA and began actively collecting dinosaur bones in Big Bend State Park from June 1938 to January 1939. Dr. Wann Langston, Jr. reports:

Dr. Richard Cifelli, Curator of Vertebrate Paleontology, Oklahoma Museum of Natural History, OU at Norman, is holding a duck-billed Kritosaurus *vertebra from Oklahoma's Texas collections.*

Dr. Stovall, at OU, hearing much inflated reports of Strain's successes, decided to send his own collectors to the area. Dr. Donald E. Savage, an OU graduate student; myself (Wann Langston, Jr.), then a high school student; and Dr. William McAnulty, borrowed from the WPA project in Oklahoma, arrived in the Big Bend in June 1938, while Strain's party was still engaged there. Over the next two months, working on some mineral claims that Strain had not acquired, we made a very modest collection of dinosaur bones compared to the tons of material recovered by Strain for the Texas College of Mines.

For a while, the most active Texas dinosaur study was being done by Oklahomans, or in William Strain's case, an Oklahoma graduate employed in Texas. Dr. Langston, who was a part of these early OU excavations in Texas, has been on the staff of the University of Texas at Austin since 1962, continuing this long-standing Oklahoma presence in Texas vertebrate paleontology. The bond between the Norman, Oklahoma collections (OU) and the University of Texas at Austin collections (Texas Memorial Museum) is still strong.

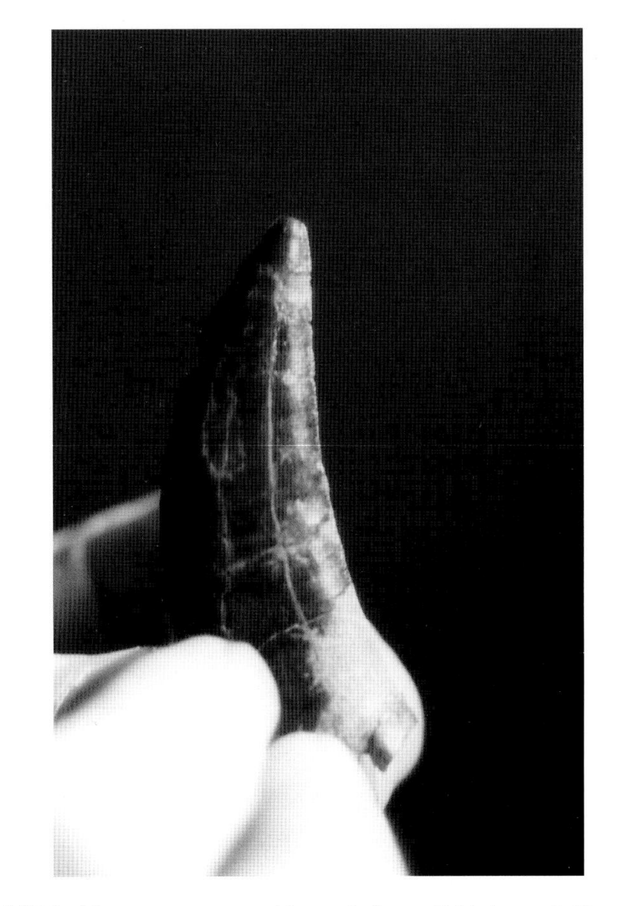

Dr. Cifelli holds a tyrannosaurid tooth from Oklahoma's Texas collections.

The American Museum

The American Museum of Natural History entered the Texas picture in roughly this same 1930s period. Contacted even prior to Strain or Stovall by ranchers and amateur collectors from the Big Bend, Dr. Barnum Brown of the American Museum and his trusted field associate Roland T. Bird were soon collecting and shipping dinosaur bones back to New York from the railroad station at Marathon, Texas, in 1940. The plaster-jacketed bones filled a box car.

Roland Bird had just finished the dinosaur track removal at Glen Rose and piled these famous tracks, each protected by plaster and burlap wrappings, into the lumberyard in Glen Rose to await later shipment to New York. Those were good days for him, I am sure. It is a glorious feeling, working in the open with people you enjoy, uncovering things that have not seen the light of day in eons of time. You have entered a chain of events millions of years in the making. Those are life-fulfilling moments for paleontologists, in the 1940s as well as today.

Bird then drove west to the Gage Hotel in Marathon to help Dr. Brown (who never drove a car) retrieve a huge six-foot-long skull of the dinosaur-age crocodile, *Deinosuchus*, down in the Big Bend.

At that time, the Gage Hotel, which is still in operation today, was the host to the American Museum paleontologists in the late 1930s and early 1940s. It was out of this same town (Marathon, Texas) and the same hotel that Dr. G. Arthur Cooper, of the Smithsonian Institution, later shipped 50 tons of Glass Mountain rock back to Washington, D.C. He cleverly freed wonderfully preserved seashell fossils of the Permian period, using a bath of acid as a gentle excavator.

UT at Austin

During the last half century, the University of Texas at Austin has led vertebrate fossil collecting in the Big Bend. They have found duck-billed dinosaurs like *Kritosaurus* and a look-alike for *Edmontosaurus,* horned dinosaurs like *Chasmosaurus* and *Torosaurus,* meat-eaters of various sizes, maybe even *Tyrannosaurus,* and also *Alamosaurus,* the last of the giant sauropods on earth.

Besides dinosaurs, huge dinosaur-age crocodilians such as *Deinosuchus* and flying giants like the pterosaur, *Quetzalcoatlus,* have been collected by UT Austin paleontologists excavating in the Big Bend. The budget for this work has been small, and some of the material still awaits

Quetzalcoatlus, *a huge aerial cousin of the dinosaurs, was a flying reptile that once soared over the Texas Big Bend.*

Dr. Ernest L. Lundelius, Jr., retired director of the Vertebrate Paleontology Lab, UT Austin, showing Henry de Mauriac, Dallas Paleontological Society, some unopened plaster field jackets yet to be examined and prepared.

preparation and study. Inside unopened field jackets are the partial fossil remains of the largest Texas dinosaur of all, a 70-foot-long sauropod *Alamosaurus.*

Dinosaur skeletons from Texas are just beginning to appear in museum exhibit galleries. Getting Texas dinosaurs on public display is one of the big challenges for museums in the near future. Much of the Texas material consists of incomplete, partial skeletons. Such reconstructions take more money and more work to complete. Still, a Texas reconstruction using partial material and lots of plaster is always more interesting to Texans than something more easily and cheaply purchased from somewhere else. Texas, like the mighty dinosaur, has a powerful following.

The First Reconstructed Texas Dinosaur

In 1962, a field collector for Director Bob Slaughter of Southern Methodist University's Shuler Museum of Paleontology found some large fossil bones in a ravine not far from Alvord in Wise County, Texas. As was often his habit, since his museum at SMU had few exhibit areas, he contacted the Dallas Museum of Natural History. He had a feeling that this was going to be a noteworthy dinosaur. Further digging by both museum crews showed that to be the case.

Dr. Langston of the University of Texas at Austin joined the excavation at Slaughter's invitation and supervised the final removal of the specimen. It was a medium-sized, plant-eating dinosaur that most resembled the Jurassic *Camp-*

Tenontosaurus *excavation in Wise County, Texas, 1962. This was a joint excavation of the Dallas Museum of Natural History, SMU, and UT Austin. This specimen was from the Lower Cretaceous, Antlers Formation.*

tosaurus. The rocks in Wise County, however, were Cretaceous in age, more than 20 million years younger than those of the Jurassic period, which yielded classic camptosaur remains. Like other paleontologists, Langston suspected these Cretaceous plant-eaters were not true camptosaurs, but there was no closer identification at the time. An identical Cretaceous specimen from Montana was on display in the American Museum of Natural History in New York, labeled as a camptosaur, and remained so labeled for some time. In 1970, however, such dinosaurs were finally recognized as different from the true camptosaurs and were given their present name, *Tenontosaurus,* by Dr. John Ostrom of Yale University.

Tentative identifications are quite proper in a young and growing science like the study of dinosaurs. Early identifications are sometimes based on incomplete skeletons or just a few bones or teeth. Young people entering this science must be prepared to do the best they can with what is known. One must often face the public question, "What exactly is that?," with the inner knowledge that one doesn't exactly know. That's the name and nature of the game.

Dr. Tom Lehman, of Texas Tech University, has commented on the habit of paleontologists to attach a name to partial Texas specimens (often just a bone or a tooth) drawn from better known, more complete specimens from up north. This assumes that dinosaur species are the same across great distances. It supposes that a given species wandered a very long way. Dr. Lehman's own research indicates that dinosaurs stuck closer to home, with limited migration. Later discoveries often find that our Texas material is, in fact, unique. That is why Texas "dinosaurol-

Bill Lowe and Arlene Pike working on replica tail vertebrae for the Wise County Tenontosaurus *mount, Dallas Museum of Natural History.*

ogists" hedge so much when asked about a particular genus or species identification. Just about all our Texas Latin dinosaur names should be prefaced with "something like a."

Lloyd Hill, fossil preparator at the Dallas Museum of Natural History during the 1980s and early 1990s. He worked on a Tylosaurus *mosasaur, a Trinity River mammoth, a large* Protostega *sea turtle, and a* Tenontosaurus *dinosaur.*

The Wise County specimen of *Tenontosaurus* went to the collections of the University of Texas at Austin, as reward for Dr. Langston's efforts. Many years later, arrangements were made through my curatorial office at the Dallas Museum of Natural History to have it loaned back to the Dallas Museum in return for its complete restoration and exhibit. In exchange, the Dallas Museum made a loan to the University of Texas of a very interesting fossil crocodile we had excavated from dinosaur-age rock, south of the Dallas-Ft. Worth Airport.

As curator of earth sciences for the Dallas Museum of Natural History, I organized a cadre of volunteers to do certain parts of the tenontosaur reconstruction under Dr. Langston's close consultation. A lot of the technical work was done in Austin's Vertebrate Paleontology Lab with Kyle L. Davies' excellent help.

My fossil preparator at that time, Lloyd Hill, was also very much involved in the project. Lloyd Hill spent many years as an amateur collector before he became a mainstay of the work on the many large skeletons we put on exhibit at the Dallas Museum of Natural History. Unperturbable, Lloyd would spend hours meticulously piecing back together a fragmented bone. He was great with jigsaw puzzles, another important ingredient in working with fossils.

A skeletal restoration by Dr. Wann Langston, Jr. of the Wise County Tenontosaurus, *currently on exhibit at the Dallas Museum of Natural History.*

The tenontosaur went on exhibit in 1989 as the first Texas dinosaur ever fully reconstructed. It stands in the Dallas Museum of Natural History for all to see. Donations from Oryx Energy Company (then Sun Oil) were indispensable in this effort.

A Proliferation of Tenontosaurs

Although the Wise County *Tenontosaurus* at the Dallas Museum was the first reconstructed Texas dinosaur, several other tenontosaurs have been found in north Texas and Oklahoma.

One of the very first tenontosaurs discovered in Texas was found by Mr. Wolfe of the Wolfe Plant Nurseries fame. He discovered it in either Hood or Erath counties in the 1930s. Mr. Wolfe donated the tenontosaur bones to a small Texas college, but their whereabouts are presently unknown.

A complete tenontosaur skeleton, minus a tail, was discovered in 1974 by Dr. Langston and his graduate student Dr. Marc Gallup in Wise County, Texas. This specimen still rests in its field jackets at the Vertebrate Paleontology Laboratory in Austin because of a lack of funds to prepare it.

A much publicized tenontosaur discovery was reported to the Fort Worth Museum of Science and History from the Doss Ranch west of Fort Worth. In 1988, Ted and Thad Williams, father and son, came across the skull of a tenontosaur and took it to Jim Diffily at the Fort Worth Museum of Science and History. I had a chance to see these fossils at the Dallas Museum of Natural History very early on, thanks to Jim. They certainly looked much like the tenontosaur we were reconstructing at the time in Dallas. This discovery, soon involving Southern Methodist University, led to a second tenontosaur discovery nearby by Gary Spaulding. Spaulding is an honored member of the Dallas Paleontological Society and a middle school

Ted and Thad Williams, father and son, with the dinosaur they discovered, now called Tenontosaurus dossi *after the ranch where it was found. On exhibit at the Fort Worth Museum of Science and History.*

teacher. His zeal to obtain classroom specimens for his students led him to find a dinosaur.

The end result, after SMU study, was possibly *two* different *Tenontosaurus* species. The analysis of the small front teeth on the Doss Ranch specimen also allowed the genus *Tenontosaurus* to be placed with some certainty in the hypsilophodont family, which had been somewhat debatable prior to that. Texas is greatly expanding the knowledge of tenontosaurs. The original Doss Ranch find (actually of two animals) has produced a mounted skeleton on display in the Fort Worth Museum of Science and History. Dr. Louis Jacobs realized that the animal now on exhibit in Fort Worth is in fact a new species, *Tenontosaurus dossi*. Jacobs believes Texas is an area where *Tenontosaurus dossi* may have evolved over several million years into a later species, which may or may not be the formerly recognized northern U.S. species *Tenontosaurus tilletti*. Texas seems to be good ground for studying tenontosaur evolution.

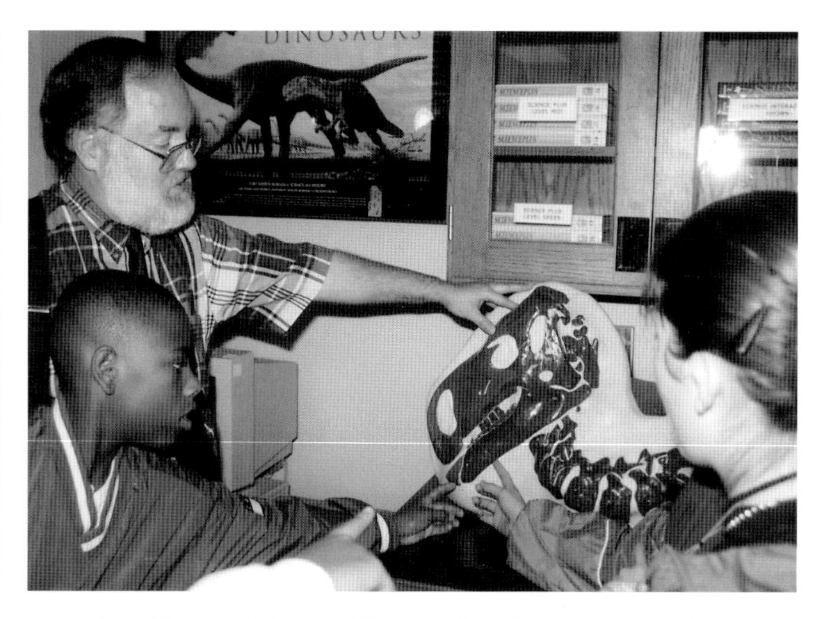

Gary Spaulding with his middle school students, showing the replica he cast of the skull and neck of the tenontosaur he found in Parker County, Texas. He uses the cast with his students while the real fossils are kept safe.

The specimen from Wise County on display at the Dallas Museum of Natural History needs further study to relate it to the two species found in Parker County by the Williams and Gary Spaulding. The Wise County animal was found farther north in Texas where the Twin Mountains and Paluxy Formations meld into the one undivided Antlers Formation. Thus it is difficult to ascribe it to the same rock formation as either the Doss Ranch tenontosaurs in the Twin Mountains Formation or to Gary Spaulding's tenontosaur from perhaps the Glen Rose Formation or the Upper Twin Mountains Formation. It may be either older or younger than the other tenontosaurs. In addition, the skull of the Wise County specimen is so incomplete as to make identifi-

Tenontosaurus dossi, *as mounted for exhibit by retired Smithsonian Institution fossil preparator Arnold Lewis. On exhibit at the Fort Worth Museum of Science and History.*

cation of the species difficult by that means. A thorough examination of the post cranial anatomy (all the bones other than the skull) is needed.

Dr. Langston believes the Wise County tenontosaur may be identical to the Spaulding specimen, and probably not *Tenontosaurus dossi.* However, everyone is reluctant, without further study, to peg any of these specimens too surely to the classic species from Montana *Tenontosaurus tilletti.* Too much geography lies in between. There is a Texas dinosaur identification job for someone. There are many such questions in Texas.

One Dinosaur Leads to Another

News of these discoveries and contact with other ranchers led to SMU getting a report of a nearby meat-eating dinosaur in the North Texas rocks, probably an *Acrocanthosaurus.* It seems Philip Hobson's family had been living with the knowledge that they had some interesting fossil bones on their ranch for some time. The other excavations in their Parker County area caused them to mention it casually to Louis Jacobs, of SMU, at a ranchers' barbecue. While work progresses at the Shuler Museum on that skeleton, the Fort Worth Museum of Science and History has placed on its lawn a concrete replica of this ferocious dinosaur as well as a replica of the plant-eating *Tenontosaurus.*

SMU and the Fort Worth Museum are presently excavating a "graveyard" of huge Lower Cretaceous sauropod dinosaurs, the big "brontosaur" kind. When a skeleton of

A sculptor's version of the plant-eating dinosaur, Tenontosaurus, *Fort Worth Museum of Science and History.*

A sculptor's version of the Lower Cretaceous dinosaur, Acrocanthosaurus, *Fort Worth Museum of Science and History.*

one of these creatures, a member of the brachiosaur family, is assembled and on exhibit, it could be 60 feet long. All this is taking place in Hood County, Texas. The creature is known as *Pleurocoelus.* It is likely to be the same kind of dinosaur that left the big four-toed bathtub footprints near Glen Rose, Texas (and elsewhere), about 115 million years ago. It will grace the halls of the Fort Worth Museum of Science and History. Fort Worth will have the distinction of showing its visitors Texas' best display of "home-dug" Texas dinosaurs, thanks in no small part to the considerable effort of the Shuler Museum of Paleontology at Southern Methodist University. The Fort Worth Museum has the potential to become a powerful leader in Texas dinosaur study and exhibit in the next century.

The SMU Shuler Museum, it should be said, is not at present a display museum. It exists for scientific study and

This Acrocanthosaurus' *name means "high spined," referring to its vertebral spines. (Sculptor's version, Fort Worth Museum of Science and History.)*

Jim Diffily, Fort Worth Museum of Science and History, and Dr. Phillip Murry, Tarleton State University, at work on rock containing fossils of the giant brachiosaurid dinosaur, Pleurocoelus. Photo by Dr. Murry.

keeps a well-managed collection for scientific reference. So it is natural that it would often help the Dallas Museum of Natural History and the Fort Worth Museum of Science and History with exhibit projects in their public halls. That has been true from its earlier director Bob Slaughter to the excellent stewardship today of Drs. Louis and Bonnie Jacobs and Drs. Dale and Alisa Winkler and a host of students. The Shuler Museum has been especially active in training future paleontologists from developing countries around the world.

Amal A. Mohamed, SMU student and fossil technician from Sudan, using a small airtool to remove matrix from bones of the giant Texas dinosaur Pleurocoelus. *SMU has been actively educating foreign paleontology students.*

A Herd of Dinosaurs

A group of volunteers worked late one night in the old basement fossil lab at the Dallas Museum of Natural History helping me put together a prehistoric elephant skeleton (found in Dallas on the Trinity River). One of them leaked

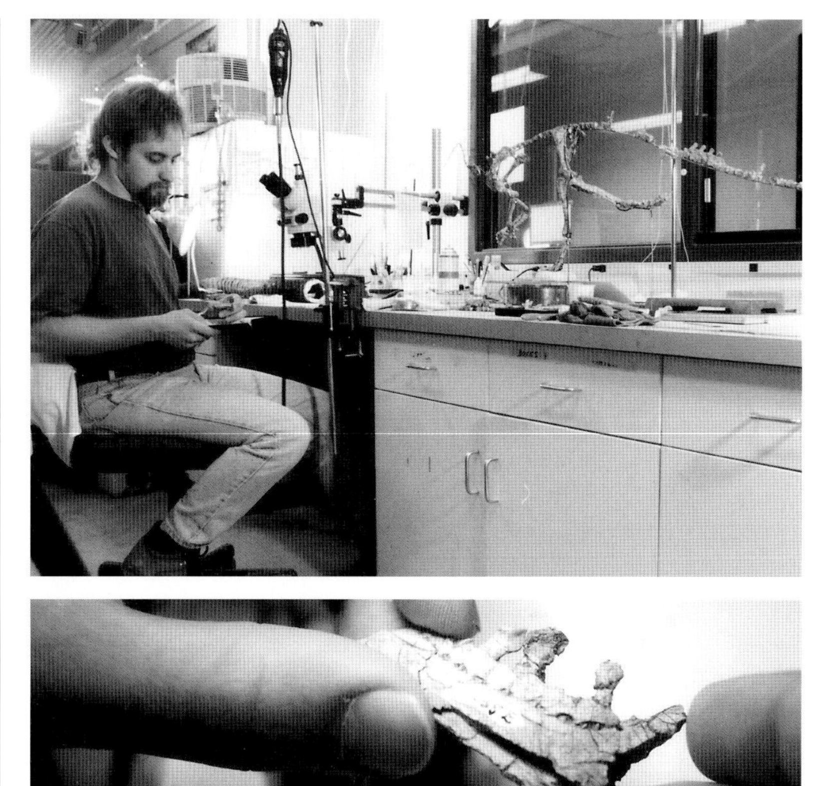

Geb Bennett, Dallas Museum of Natural History fossil preparator, at work on the first mounted Proctor Lake dinosaur, now on temporary exhibit in the museum. This is a medium-sized specimen for the site.

Mounted skeleton of a small Proctor Lake dinosaur, Dallas Museum of Natural History.

Dr. Dale Winkler, SMU, with foot of a Proctor Lake "bird foot" ornithopod dinosaur. The site produced many skeletons.

the word that SMU was working on a huge bone-bed of small, but nearly complete, dinosaurs out near Stephenville, Texas. This site, found by a Tarleton State University student, Rusty Branch, is perhaps the most remarkable Texas dinosaur bed ever. It was a nesting ground of medium-sized and infant plant-eating dinosaurs called (on first research) only by the family name, hypsilophodont. There were so many skeletons around Proctor Lake that security was a big issue. A federal permit is required to collect in such an area, with stiff penalties for violations. It is guarded today by the U.S. Army Corps of Engineers. Although no eggs have yet been found, the presence of very young individuals of several age groups gave rise to the thought that this was further proof that some dinosaurs lived in small family groups

Close-up of Proctor Lake dinosaur's foot. Like other ornithopod dinosaurs, these small creatures were plant eaters and probably lived in large herd communities.

for a while after the young were hatched. Reptiles are usually quite unattentive of their young after hatching, but other dinosaur sites in Montana had shown evidence that some mother dinosaurs tended their young offspring over a period of years. Proctor Lake has only begun to be studied.

SMU is now committed to assist Geb Bennett, the present fossil preparator at the Dallas Museum of Natural History, to reconstruct a few of these nesting dinosaurs for a future Dallas Museum of Natural History exhibit, with help from a generous grant by former Texas Governor, William P. Clements.

Especially worthy of mention is Dr. Dale Winkler's work at SMU with these Proctor Lake dinosaurs. He has been the major liaison with the Dallas Museum of Natural History in preparing the skeletons for future exhibit, and he will be instrumental in determining their final identification. As with most things Texan, they will definitely turn out to be quite unique.

Dinosaur Preparators

At the time of the preparation of several of the Dallas Museum fossils, several other (student and post-graduate) preparators took on prominent roles, mostly in the J. J. Pickle Research Campus (formerly the Balcones Research Center) in Austin. Hardly a project there failed to involve Kyle Davies. Kyle, who is currently a fossil exhibit technician at the Oklahoma Museum of Natural History in Norman, is a paragon of meticulous and accurate fossil work. He worked on many projects for the Dallas Museum under Dr. Langston's guidance, including the Rockwall County,

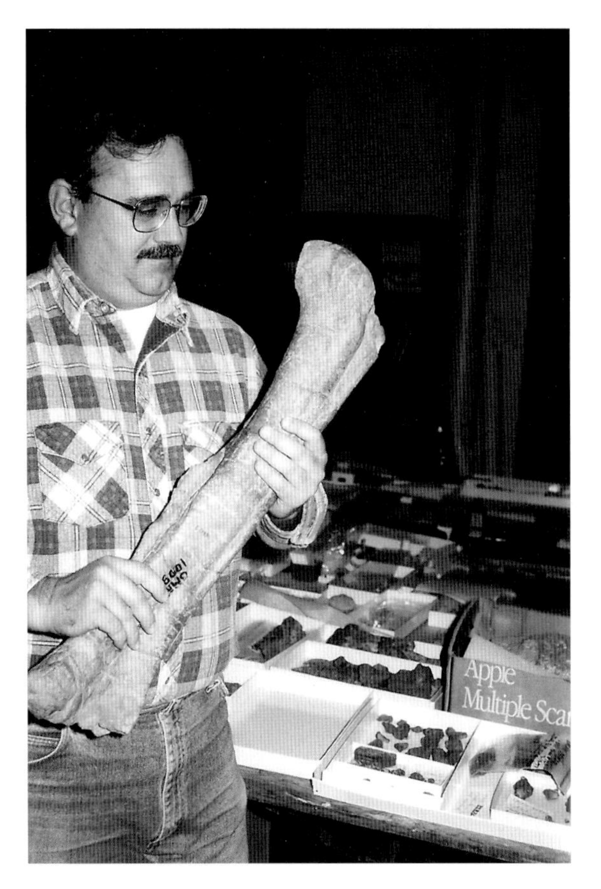

Kyle L. Davies with a leg bone of the duck-billed dinosaur Kritosaurus. *Now a fossil technician at the Oklahoma Museum of Natural History, his master's thesis at UT Austin was about Big Bend duck-billed dinosaurs.*

Earl Yarmer, UT Austin Vertebrate Paleontology Lab, examining toe bones of a duck-billed dinosaur from the Texas Big Bend.

Heath mosasaur *Tylosaurus proriger* and the Wise County tenontosaur. Kyle once helped us move the whole 30-foot-long mosasaur, steel framework and all, from one floor up to the other. This difficult chore later occasioned Davies and Langston to put permanent wheels on some later mounts. One such mobilized mount was our Wise County tenontosaur. A dinosaur on roller skates! Innovative engineering is definitely a part of dinosaur work.

No discussion of skilled and important Texas fossil preparators would be complete without mention of Earl Yarmer and Bob Rainey, who have worked for many years with Drs. Wilson, Langston, Lundelius, and Rowe at the UT Vertebrate Paleontology Laboratory in Austin. Earl Yarmer's father and grandfather were fossil preparators at the great Carnegie Museum in Pittsburg.

Always knee-deep in something at the UT Austin lab, one of the most important Texas dinosaur projects Yarmer and Rainey have recently undertaken is the reconstruction of a magnificent frilled skull of the horned dinosaur *Chasmosaurus*. It was found in the Big Bend in 1991. Dr. Catherine Forster, now at the State University of New York at Stony Brook, has been central to its excavation, scientific study, and publication (Forster, Sereno, Evans and Rowe).

What Does a Fossil Preparator Do?

Fossil preparation is a combination of scientific knowledge, artistic skills, and construction techniques. It involves an understanding of bones, chemicals, molding rubber, plaster or resin casting, sculpture, clay modeling,

Frilled skull of the Texas horned dinosaur Chasmosaurus mariscalensis *still in part of its protective plaster field jacket. Such a complete frill makes this specimen rare and unique.* Photo by Dr. Catherine A. Forster.

blacksmithing, welding, and woodworking. Basically, it is anything it takes to take an incomplete fossil skeleton of a creature and make it durable and complete. It is a fascinating occupation, although seldom a good-paying one. Reward is in the satisfaction of knowing that one's work is for the benefit of ages to come. Texas has far fewer full-time professional preparators than paleontologists. Volunteers help fill the gap but can never replace highly trained, experienced professional preparators.

A Dinosaur Welder?

One should not overlook the important contributions of different kinds of workers in putting together fossil skeletons for museum exhibit. A very artistic welder is nearly as important to a fossil reconstruction as is the paleontologist. At the time of construction of the huge *Diplodocus* skeleton at the Houston Museum of Natural Science, Dr. Langston, who was in charge of the project, came across the unique skills of art welder and exhibit designer John Barber. John has helped Langston reconstruct not only the Houston *Diplodocus* (Texas' largest mounted dinosaur), but also a large mosasaur, sea turtle, and the *Tenontosaurus* dinosaur at the Dallas Museum of Natural History. John also contributed his expertise to the growing number of fossil exhibits at the Houston Museum of Natural Science. Without John Barber's skill, many of the Dallas Museum's fossils wouldn't have a leg (support) to stand on.

John Barber, Houston, Texas, welding the Wise County Tenontosaurus *at the Dallas Museum of Natural History.*

John Barber, shown with his metal framework for the Dallas Museum's huge fossil sea turtle. With Dr. Langston's supervision, Barber did the framework for the 70-foot-long Diplodocus *at the Houston Museum of Natural Science. A welder-artist!*

It's a Long, Hard Job

The very long time and tedious work necessary to excavate, prepare, study, and assemble a fossil skeleton always surprises people. At many excavation sites, a common question is, "When will this be on exhibit in the museum?" I often pick out a nearby, very young child and say, "When she is in high school!"

Removing the rock, called *matrix*, from a delicate fossil bone is a painstaking process. Missing parts must be sculpted in clay and cast in plaster or fiberglass. To support the skeleton with a steel framework requires careful planning. The Hood County sauropod *Pleurocoelus* and the Parker County *Acrocanthosaurus*, both mentioned earlier, will be a slow, painstaking job because of the dense matrix. But, Texas needs to see a standing skeleton of one of its big sauropod dinosaurs.

I remember on one hard rock site, Arlene Pike, my field associate, proclaimed loudly, as her chisel was replaced by her thumb, under the hammer, "This isn't digging, it's stone carving!" She was right.

More Dinosaur Workers Than Ever

The Dallas Museum of Natural History as well as SMU, the Fort Worth Museum of Science and History, and the Strecker Museum at Baylor University have introduced new life into the study of Texas fossils. Both SMU and the

Dallas Museum of Natural History have fully functioning fossil laboratories in north Texas. This should take some of the pressure off the state's oldest and still extremely competent vertebrate fossil laboratory at the University of Texas in Austin.

The number of amateur and professional discoveries in north and central Texas, the Big Bend, and in the Triassic rock of west-central Texas, have kept all the Texas laboratories very busy. During my time there, the Dallas Museum of Natural History often had to act as a place of last resort for fossil finds that the other overworked institutions did not have time to handle. Sometimes discoveries are referred from one institution to another. It is frustrating for an amateur fossil hunter or a landowner to make a fine discovery, and perhaps even have the good sense to protect it temporarily from harm, only to be told that all the professional paleontologists are too busy. This is something Texas paleontologists must try to avoid happening.

Remember the plaster field casts sitting unopened for decades after their discovery in virtually every Texas fossil laboratory? There is great need for benefactors to come to the aid of museums and universities with money for fossil excavation, study, and exhibit. What better place to put one's name than on a Texas dinosaur exhibit?

One Man's Struggle to Reconstruct a Dinosaur

Let me tell you a story about a man and a dinosaur. At a symposium in the mid 1970s where we were both speakers, I first met the late Professor John Brand of Texas Tech University. Dr. Brand was an invertebrate paleontologist, very famous for his work with Texas sea urchins. Texas Tech had purchased an allosaur skeleton (the large Jurassic meat-eater) from the University of Utah, many years earlier. Everyone at Texas Tech seemed too busy with other worthy projects to put it together, and there was never enough money. Finally, John Brand decided to take on this vertebrate "urchin." Sea urchins were in fact his thing. He was determined to put that dinosaur together. To help support the project, he sold "Take an Allosaur to Lunch" belt buckles to local business groups.

The first obstacle he and his student helpers encountered was that the bones were not from *one* individual. They were from several animals of varying sizes. The leg bones, for instance, were of different lengths. That was finally straightened out by the experts at Salt Lake City who had collected the bones. Because fossil skeletons are never found in a complete state of preservation, skeletons in museum displays are often composites of the bones of several individuals.

Many of the bones they had were deformed by pressure underground. That can happen during the process of fossilization or in the millions of years of burial—150 million years of burial in the rock is a harsh experience. It was obvious to John Brand that what would result by using the real fossils would be a deformed-looking animal. Only by making clay replicas of each deformed fossil and attempting to decipher its original shape were Dr. Brand and his students able to proceed. Plaster casts were made from the clay replicas.

The final mount is on exhibit in The Museum at Texas Tech. This story illustrates some of the many problems anyone faces when attempting to make fossils "live again" in museum exhibits. There is much more to it than just digging up some fossils and rigging some rods to hold 'em up. The entire process is a sea of technical problems. The best solution for keeping some sanity in the process is to take one problem at a time. Future problems will reveal their solutions by the time one gets to them.

Texas Tech Steps Up Its Dinosaur Work

Today at Texas Tech University, Dr. Sankar Chatterjee labors to reap a bountiful harvest of Triassic fossils from a little hillside near Post, Texas.

Dr. Chatterjee has excavated an impressive list of creatures from these Texas Triassic rocks. One of his most physically impressive creatures was a large meat-eater with a skull reminiscent of the toothy skull of *Tyrannosaurus*. He named it *Postosuchus*, after Post, Texas, near the site where it was found. It was a large predator, with its own version of the upright stance usually associated with dinosaurs. After further study, *Postosuchus* has instead been assigned to a group of large, non-dinosaurian carnivorous reptiles. Such creatures were for a time undoubtedly major competitors for dominance with the early dinosaurs.

Dr. Chatterjee's discovery of *Shuvosaurus*, a small, light-boned creature of ostrich-like proportions, has much to suggest it was an early dinosaur. Ironically, contrary to what we would expect of early dinosaurs, it was in no way primitive, but was rather highly evolved, somewhat resembling the so-called ostrich-mimics of much later times. To

Postosuchus *with a dead* Placerias. *These were some of the dinosaurs' Triassic competitors.*

Dr. Sankar Chatterjee in the collection area of The Museum at Texas Tech University with a large toothy skull of one of the Triassic dinosaurs' fiercest competitors, the rauisuchian reptile called Postosuchus.

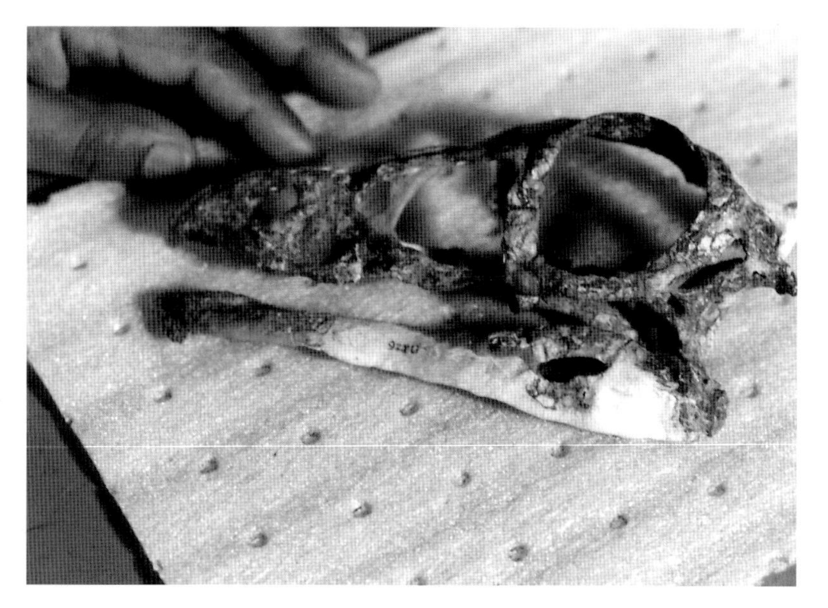

The delicate skull of the dinosaur Shuvosaurus. *The refined and bird-like characteristics of this specimen make it very unusual for the early time period of the Texas Triassic.*

occur so early in the dinosaur age and yet resemble bird-mimic dinosaurs, which are considered much later in dinosaurian development, only shows that more study is needed to understand *Shuvosaurus'* place in dinosaur evolution. *Shuvosaurus inexpectatus* is well named by Dr. Chatterjee, in that such a dinosaur is not to be expected in rock as old as the Triassic. The accompanying photos show its delicate and refined skull—a skull devoid of teeth. Dr. Chatterjee argues that this skull is that of a theropod dinosaur. On the other hand, Long and Murry, in their recent work on the Triassic period, offer some evidence for it being a rauisuchian, and not actually a dinosaur. The rauisuchians were a varied group of dinosaur competitors that included some species that somewhat resembled dinosaurs. Such differences of opinion are common when dealing with the fragmentary and often unique Texas Triassic dinosaur material.

Other Triassic Work

The University of Texas at Austin did a lot of Triassic period work in Howard County, especially in the vicinity of Otis Chalk, a place named after a local rancher-landowner. Dr. Wann Langston, Jr. was very instrumental in the discovery there of many non-dinosaurian reptiles and also Triassic amphibians. In more recent years, Dr. Phillip Murry, now of Tarleton State University, extensively worked these beds, which are not far from Big Springs, Texas.

In an interesting conjunction of efforts old and new, Dr. Murry noticed a very small femur (thigh bone) head in the cataloged collections at the UT Austin lab. It was the angled, ball-tipped part of the femur that, very characteristic of dinosaurs, fits into the pelvis (hip) with a distinctive, almost 90 degree angle. That makes folks like Dr. Murry pay attention. This little bone fragment is a close look-a-like for the femur head of a herrerasaurid, one of

Femur head of possibly Texas' oldest known dinosaur, Chindesaurus bryansmalli *(TMM 31100-523). Most identifications of Texas Triassic dinosaurs are educated speculations based on small, scattered, bone fragments.*

the very earliest known theropod dinosaurs, usually found in Argentina. It quickly became one of the most important little bone fragments in Texas. Further study and a pelvis collected even earlier by Dr. E. C. Case, of Michigan, confirmed the naming of a new Texas dinosaur, *Chindesaurus* by Drs. Robert A. Long and Phillip A. Murry (1995). An

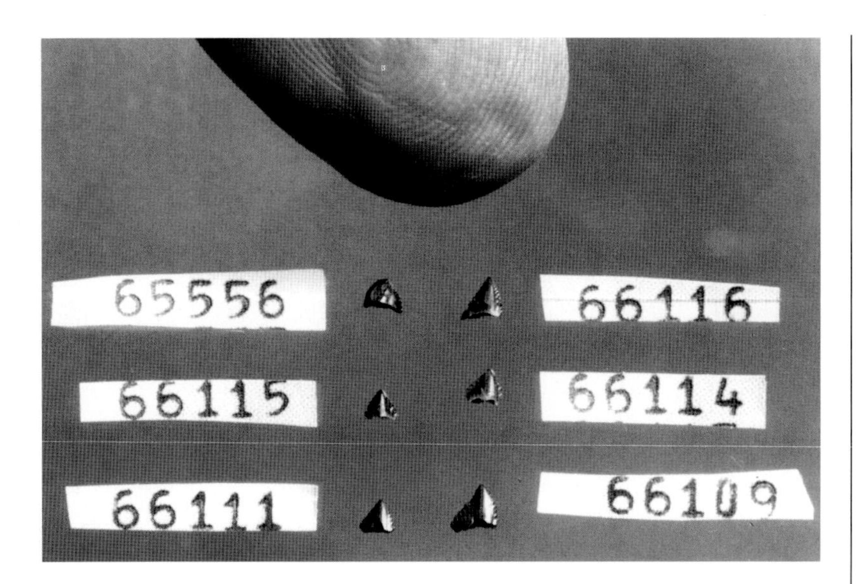

Teeth of Tecovasaurus murryi, *a small plant-eating dinosaur of the Texas Triassic period. It is amazing that just small teeth like these can provide evidence of a dinosaur species. Even smaller teeth of dinosaurs are found in Texas rock.*

interesting sidelight is that this small fragment may also be the oldest known piece of a Texas dinosaur.

From Triassic deposits in the Texas Panhandle, Dr. Phillip Murry collected what must be the sparsest evidence of a particular Texas dinosaur, *Tecovasaurus murryi.* It is known from only the small collection of very tiny teeth. When I came to Tarleton State University to photograph them, he asked me if I was ready for such small dinosaur fossils.

To me, that is what makes them fascinating. Dr. Murry hopes that in the future much more will be found of *Tecovasaurus murryi.*

New Hunters in the Big Bend

Dr. James Carter, geosciences professor at the University of Texas at Dallas, is interested in everything to do with rocks, minerals, and fossils. He has long entertained the thought of a museum on his university campus, just north of Dallas. On a visit I made to UTD in the winter of 1995, he asked if I would like to see their "dinosaur." Off we went, into one of the laboratory areas. There on the tables lay much of the skeleton of a juvenile *Alamosaurus,* the potentially biggest species of Big Bend dinosaur. It was discovered on a hiking reconnaissance of the Javelina Formation by UTD and now SMU geology student Dana Biasatti. What a way to start your geological career!

Teams from UTD and the Dallas Museum of Natural History led by Dr. Homer Montgomery and Dr. Antony Fiorillo have already excavated parts of three alamosaurs in

Dr. Homer Montgomery, UT Dallas, is part of a joint effort with the Dallas Museum of Natural History to study and excavate dinosaur fossils in the Texas Big Bend. Shown are Alamosaurus *bones in the ground.* Photo courtesy of Dr. Montgomery.

Dr. Montgomery's students carry bones of Alamosaurus *in protective field jackets.* Photo courtesy of Dr. Montgomery.

the Big Bend. Many other dinosaur-age fossils are in the vicinity. These excavations are being carried out under permits from the National Park Service and cooperation from the Big Bend National Park. The university and museum paleontologists are emphatic that their efforts are aimed at not only obtaining dinosaur fossils, but also leading to an understanding of the overall paleoenvironmental interpretation. In such an interpretation, even the slightest angle at

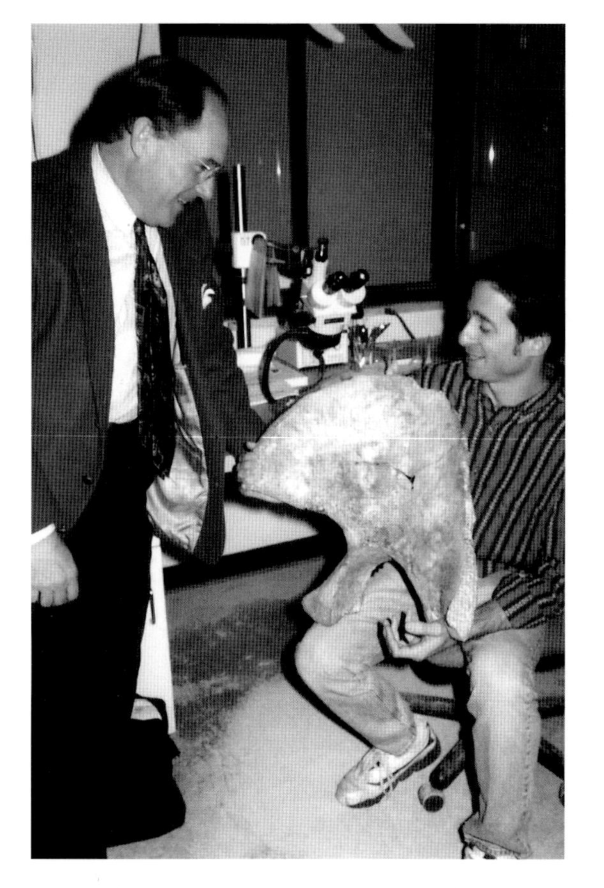

Dallas Museum of Natural History director, Robert Townsend (left) and Dr. Antony Fiorillo (right) examine part of the pelvis of a juvenile Alamosaurus *from the Texas Big Bend.*

which the fossils lie is potentially important to understanding what happened at the site. On the face, such sites resemble a gigantic jumble of usually disassociated bones, but to a trained eye and with careful measurement, they can reveal the story behind the seeming jumble. Dr. Fiorillo's specialty is *taphonomy,* the death postures and effects of burial on fossil creatures. That is another example of why, even though amateur paleontologists can be of great help (school teachers are, in fact, being trained on this site), professional paleontologists should direct most excavations. Personally, I prefer a cooperation between all types of paleontologists.

Non-Professionals Make Important Contributions

Amateurs and volunteers have always made fine contributions to the discovery, excavation, and study of dinosaurs. Several members of the Dallas Paleontological Society have come across wonderful dinosaur discoveries in the course of checking out roadcuts and rock quarries in the normal round of their fossil hunting. Such societies exist also in Austin and in Houston. In fact, almost every area of Texas has rock and mineral clubs that also take an interest in fossils. The monthly meetings of such groups and their field

Dallas Museum of Natural History volunteers with many excavations and fossil preparations to their credit: from left to right, Roberta Dierken, James (Jim) Merrett, and Charles Wyatt.

trips can be a very good way for a novice to enter this interesting avocation. Rock and mineral shows occur annually in most cities, and often specifically fossil-oriented shows occur. Check with a local rock shop for a schedule of these events and club meetings. Ordinary folks have contributed a lot to the knowledge of fossils in Texas. In my career at the Dallas Museum, I used many members of such groups as very productive volunteers. Several other Texas museums and fossil laboratories do the same. There's a way to get your foot in the door. Check around.

What One Person Can Accomplish

When I think of dinosaur discoveries and amateur collectors, I first think of my friend and frequent volunteer, Gary Byrd. He is a roofing contractor by trade. He has always liked fossils, and he majored in geology in college. As he drives around from job to job, he looks for fossil locations and checks them out when he can. In 1993 and 1994, while keeping an eye on a large construction site at a major highway intersection north of Dallas, he reported to me several noteworthy fossils, which we either excavated or he brought to the museum whole. Among this group of finds were two nearly complete mosasaur skulls. These were both species previously unknown in Texas. On that same site, he found two fish fossils that were totally new to the museum's collections. One day, as I was helping him excavate a group of small fish skeletons, a construction

Fossil discoverer Gary Byrd at the site of one of his North Texas discoveries. Responsible collectors like Gary constantly check every rock outcrop and report their discoveries to professional paleontologists. The fossil here is a rare Cretaceous shark.

worker asked Gary if we would be interested in a big rock with bones sticking out of it. Of course, we would be. It turned out to be a mass of mosasaur vertebrae, probably most of a skeleton.

Gary Byrd's Dinosaur

Then one morning, I believe it was in 1994, Gary called me to report that he had found some dinosaur-like bones along a roadside in a small rock exposure near Grapevine Lake. I asked my best helper, volunteer, and friend Bill Lowe to join Gary at the site to see what it might be. There had been a lot of dinosaur tracks and some dinosaur bones found not far from Gary's discovery. Bill Lowe came back to the museum to tell me that it looked like dinosaur bone, and lots of it. It seemed to extend back into the small bank. The Dallas Museum of Natural History could have excavated the site, but SMU had a graduate student from Korea, Yuong-Nam Lee, who was studying that particular rock formation, called the Woodbine. He was especially interested in its dinosaurs. So I called SMU and turned the site over to them. The skeleton turned out to be a very ancient duck-billed dinosaur (hadrosaur). It was one of the earliest known from North

Jason Head, SMU graduate student, with jaw bone of the hadrosaur (duckbill) he is scientifically describing as a new genus and species Protohadro byrdi. *The specimen was found by Gary Byrd in northeast Texas.*

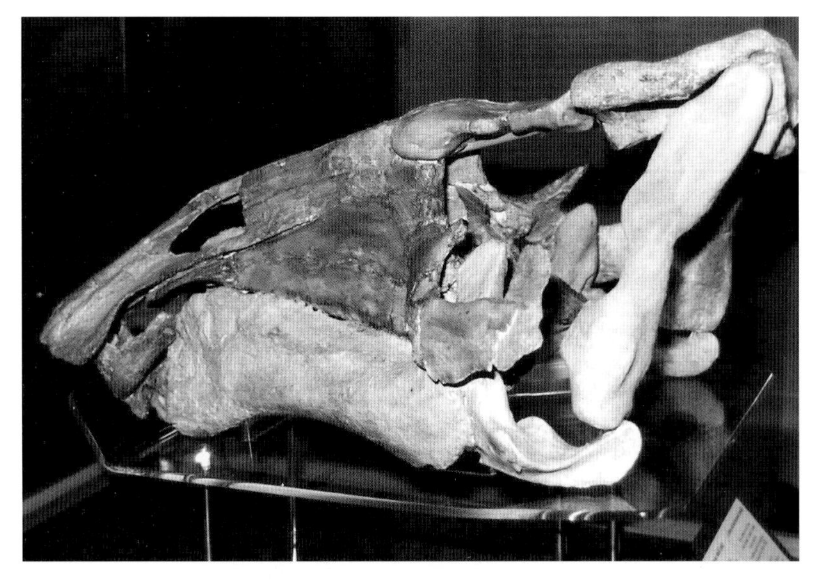

A replica of the Protohadro byrdi *skull on exhibit at the Fort Worth Museum of Science and History as part of the Lone Star Dinosaur Exhibit. It is a very primitive, early form of duck-billed dinosaur.*

America. Best of all, it had an unusually complete skull, the most important part of the skeleton one can find for study or exhibit. A replica of the skull of this duck-billed dinosaur has traveled far and wide in the Lone Star Dinosaur Exhibit, which SMU and the Fort Worth Museum of Science and History prepared and circulated very successfully throughout Texas. Yuong-Nam Lee has now published the results of his studies of Woodbine Formation dinosaurs and returned home. He told me that he can study very similar rock layers to ours — in Asia.

An SMU graduate student, Jason Head, is about to officially publish the description of this new dinosaur, which he named *Protohadro byrdi,* "Byrd's early hadrosaur."

Knotty Nodosaurs

Another example of amateur paleontologists contributing importantly to science is found in the experiences of John and Johnny Maurice, a father and son fossil-hunting duo. They particularly like to hunt a muddy-looking rock layer called the Pawpaw Formation. It formed during one of those partial and rare northeast Texas retreats of the Cretaceous sea. In 1989, at a site in Fort Worth, Johnny Maurice found some very tiny fossil bones, which were shown to Dr. Louis Jacobs at SMU. The bones were of a very young and very small nodosaur, an armored dinosaur shaped much like a big "horny toad." Evidence indicates that it was washed into the murky waters of the Pawpaw sea and was scavenged by small sharks and crabs, about 100 million years ago. Because the Maurices are careful collectors and preserve as much information as possible where they collect, Dr. Jacobs was able to put together the

John and Johnny Maurice examine fossils of North Texas nodosaurs, which they, along with Cameron Campbell and Rob Reid, were involved in finding. Described by SMU graduate student, Yuong-Nam Lee, they are a new genus and species, Pawpawsaurus campbelli, *named after Mr. Campbell.*

story of the drowning death and shark predation of this tiny armored baby dinosaur. Later, the Maurices and their friend Rob Reid were able to add bones to a similar discovery by Cameron Campbell of an adult nodosaur skull found in much the same area as the baby nodosaur find.

They found various scattered and missing pieces of the skull, until it was nearly complete. That is an excellent example of meaningful amateur collecting. As a result, we now have the nodosaur species *Pawpawsaurus campbelli.* All of these fossils are in the collections of the Shuler Museum at SMU, where they will always be available to scientists for research. I think there are more nodosaur species to come from this SMU research.

It has been my experience that most amateur collectors would prefer to be helpful to science. It would help if amateurs would put up the money or help find donors to support research on their discoveries. It is an interesting idea that paleo societies might consider. Support your local paleontologist with time and money.

Johnny Maurice takes a special interest in his own nodosaur find. The few bones on the life-sized outline, in this part of the Lone Star Dinosaurs Exhibit, are about all there is of Johnny Maurice's juvenile nodosaur. Though small and few, they are a great discovery.

Vanished Without a Trace?

One definition of a fossil mentions that they are "the remains or traces of ancient life." Most people think of dinosaur fossils as some part of the dinosaur or at least a rock replaced replica of such—a vertebra or a skull, for example. They think a dinosaur can only have as many fossils as it has parts. Not so!

Let me fantasize that I own a pet dinosaur—the plant-eating kind, of course. I then put two poles in the ground 100 feet apart and spread mud between them. I lead my pet dinosaur, step by step, through the muck, from one pole to the other. After securing the dinosaur, I can reap a harvest of potential dinosaur fossils: dinosaur footprints. The point is that one creature can make many fossils.

If my dinosaur answered the intestinal call of nature while tied to the pole, the products would be called *coprolites* (dung) fossils. Notice, I have not used up any actual part of my dinosaur. Fossil footprints and coprolites are *trace fossils* of dinosaurs. Skin imprints are another kind of trace fossil. Imprints of the surface texture of the skin of dinosaurs are occasionally discovered in the stony shroud that covers a dinosaur's skeleton. The previously mentioned *Tenontosaurus* from Wise County, Texas, left some skin impressions in the surrounding rock matrix. The imprints were collected as fossils, just as were the fossilized bones. Such trace fossils are technically termed *ichnites* and their study is called *ichnology.*

The scarcity of tail drag marks, another kind of trace fossil, has recently caused some paleontologists to reevaluate the earlier opinion that most dinosaurs dragged their tails on the ground. The presently popular idea is that

Jim Merrett with pebbly imprints of the skin of the Wise County Tenontosaurus.

These imprints of the skin of the Wise County Tenontosaurus *were discovered during the preparation of the specimen. Had they been noticed during the actual excavation, many more might have been salvaged. A useful caution to future excavations.*

dinosaurs were more active, often quick moving creatures with tails often elevated.

Texas Has Lots of Tracks

The most plentiful and popular trace fossils are footprints. Texas boasts some of the finest examples known anywhere in the world. Among such are the famous dinosaur footprints in the Glen Rose limestone of central Texas, particularly near the town of Glen Rose and the nearby Dinosaur Valley State Park.

Visitors to Texas, especially campers, should spend some time at Dinosaur Valley State Park, just west of Glen Rose, Texas, on Rt. 67. An interesting museum exhibit is housed in the entrance building, and the park has easily visible

Visitors observing trail of three-toed, carnivorous dinosaur footprints in the bed of the Paluxy River, Dinosaur Valley State Park, Glen Rose, Texas. The footprint name of such tracks is Irenesauripus glenrosensis. *The maker of the tracks is most probably the Early Cretaceous meat-eater* Acrocanthosaurus.

Exhibit gallery in entrance building of Dinosaur Valley State Park, Glen Rose, Texas. Shown on the mural is an Acrocanthosaurus *attacking a* Pleurocoelus. *This park provides an excellent short lesson in Texas dinosaurs for the public.*

dinosaur tracks in the bed of the Paluxy River. Low water times are the best. Full-service and comfortable camp sites are available.

The first Texans, traveling hunters, may have seen the giant three-toed, bird-like tracks in central Texas streambeds 20,000 years ago. They were either ecstatic over the prospect of a ten-ton meal or they were scared out of their wits. Who wants to come face to face with such a huge creature? They needn't have worried. By the time these hunters arrived in Texas, the makers of those particular tracks had been gone for over 110 million years.

Tyrannosaurus *and* Apatosaurus *models at Dinosaur Valley State Park, Glen Rose, Texas. These models are from the New York World's Fair. Both kinds come close to being Texas dinosaurs. There is a tyrannosaurid in the Big Bend and* Apatosaurus *occurs in Oklahoma.*

The Leander Lady

A few years ago a native American woman's skeleton was found near Leander, Texas, not far from the San Gabriel River. At first thought to be over 10,000 years old, it is nonetheless one of the oldest human skeletons ever found in Texas. In some newspapers, she was called the Leanderthal Lady. Around her neck may have hung a fossil shark's tooth from the Age of Dinosaurs; at least it was in her grave and on her chest when her skeleton was excavated. Not many miles away in the San Gabriel riverbed, are several very obvious dinosaur footprints. If she was a fossil collector, had she seen the dinosaur tracks nearby?

Glen Rose, Texas

Much has been written about the dinosaur tracks near Glen Rose, Texas. Specifically, the scientific bombshell at Glen Rose was the huge, bathtub-sized tracks and trails of "brontosaurs," not the more obvious bird-like dinosaur footprints anyone could see. The three-toed tracks had been recognized to be dinosaur tracks for many years. Americans first knew about three-toed dinosaur tracks, thought to be giant bird tracks, by the discovery in 1802 and publication in 1858 of drawings of tracks found plentifully in Connecticut. Dinosaurs had only been recognized as a group, Dinosauria, in 1841 by the English scientist Sir Richard Owen. For more than half a century afterwards, large three-toed fossil tracks were associated with dinosaurs. Yet, the big potholes left by large sauropod dinosaurs went unrecognized as dinosaur tracks until the late 1930s. Perhaps some unnamed old codger at Glen Rose was the first to be aware that there were two kinds of trackways in the bed of the Paluxy River, but until 1938 none had caught the attention of the scientific community. W. B. Moss, a local Glen Rose collector, considered the larger impressions to be prehistoric elephant tracks, if anything. Texas ranchers' main connection with their dinosaur tracks was to dig them up and sell them. The smaller three-toed tracks were easier to dig and brought better prices. After all, the bigger impressions were not as showy and were much more trouble to transport.

The closest Texas paleontologist to Glen Rose, Dr. Ellis Shuler at Southern Methodist University in Dallas, had examined the three-toed prints as early as 1917, and probably came very close to discovering the meaning of the bigger kind. But if Dr. Shuler realized what the larger prints were, he didn't write about them.

Eventually, Glen Rose entrepreneurs decided to add "homemade" human footprints and some enhanced dinosaur tracks to their list of wares by just outright carving them. It was these fake tracks, human and dinosaur, that came to the attention of American Museum paleontologist Roland T. Bird as he searched for fossils through the West in 1938. His boss, Dr. Barnum Brown in New York, asked him to investigate some strange impressions in the bed of the Purgatory River in southeastern Colorado. There he saw what may have been an eroded trail of giant sauropod prints, but they were past recognition. Intrigued, however, by some fake tracks, supposedly from Glen Rose, Texas, that he found in a Gallup, New Mexico, trading post, Bird took a southern route through Texas on his journey back to New York. There in Glen Rose, local common curiosity for three generations and science came together. Bird returned to New York with a smile on his face, having recognized in the Paluxy riverbed the world's first undeniable "brontosaur" (at least, sauropod) tracks. R. T. Bird's detailed story of that discovery is wonderfully told in his book, *Bones for Barnum Brown: Adventures of a Dinosaur Hunter,* published in 1985.

After some unsuccessful attempts to raise money for an expedition to excavate the Texas dinosaur tracks, Sinclair Oil Company came to the rescue. In 1940, R. T. Bird came to Texas to engineer the extremely arduous removal of a great trail of Glen Rose dinosaur tracks. This trail contains tracks of two kinds of dinosaurs: one the huge "brontosaur" (which we today call *Pleurocoelus* or by its "track name" of *Brontopodus)* and a good-sized meat-eater (most likely *Acrocanthosaurus* or, at least, the "track name" of *Irenesauripus).* Bones of both dinosaurs have been found in the Glen Rose vicinity.

Dr. Langston's Special Perspective

Dr. Langston comments on the Glen Rose tracks:

It is not usually possible to say with certainty that any given track was made by any particular species of dinosaur. Identifying features are not abundant on the soles of feet or the bottoms of toes. So identifications are usually based on the most general of features. The best that can usually be said is that a track was most likely made by a duck-billed dinosaur, a carnivorous dinosaur, or a sauropod, etc.

Of the untold number of sauropod species that existed, two features of the fore and hind feet are almost universal: the "hands" had five fingers, but only the "thumb" bore a claw. The other digits were blunt-ended like the "fingers" of an elephant. The hind foot bore claws on only the first three of its five toes. Such characters are often reflected in sauropod footprints. The big brontosaur tracks from the Glen Rose Formation are distinctly different in that the hand had no claw, whereas, the hind foot had four.

In 1950, a field party from Chicago's Field Museum, searching for Lower Cretaceous mammal remains in the Paluxy Formation of Wise County, Texas, came upon the left hind foot of a sauropod dinosaur standing almost upright in the light gray "packsand" as though the animal had been trapped in quicksand and buried standing up! When this specimen was freed from its rocky matrix, its foot was found to have *four* claws. Thus there is more reason than usual to claim, as does Dr. Mark Gallup (a former graduate student at UT Austin), that the famous four-clawed Glen Rose sauropod footprints were made by the same species of dinosaur to which the Wise County foot belonged. Unfortunately, we do not know details of the hind foot skeleton of *Pleurocoelus,* so we cannot say for sure that the Field Museum's specimen from Wise County belonged to that particular sauropod. It seems a good bet, however, because the only sauropod so far recognized from bones and teeth in the Lower Cretaceous of Texas is *Pleurocoelus.*

In 1989, the Glen Rose sauropod tracks were given the formal name of *Brontopodus birdi* by three experts in the field of dinosaur tracks, Drs. James O. Farlow, Jeffrey G. Pittman, and Mr. J. Michael Hawthorne. For the time being, this name can be used to designate all the other sauropod tracks currently known from Texas and Arkansas. These gentlemen concluded that whatever animal actually made these tracks, *Pleurocoelus* is a good candidate.

Pleurocoelus was related to the group of mainly Jurassic sauropods known as brachiosaurs. Whether the *Brontopodus track-maker* had the long "arms" and neck of the more typical *Brachiosaurus,* as seen in the movie *Jurassic*

Park, remains to be determined by future discoveries. *Pleurocoelus* did have longer than usual forelimbs, and the circumstantial evidence seems compelling.

Bird's Glen Rose Trackway

A trackway of Glen Rose tracks excavated by R. T. Bird eventually came to exhibit under the skeletons of two unrelated and much older Jurassic dinosaurs in the American Museum. Several tracks from the same trackway are displayed at the Texas Memorial Museum, University of Texas at Austin. Several additional specimens were donated to Southern Methodist University, Baylor University, Texas A&M, the Smithsonian Institution, and Brooklyn College.

One of Bird's very large *Brontopodus* tracks, first given to Dr. Shuler at SMU, is now on display at the Dallas Museum of Natural History. The track now at the Dallas Museum of Natural History was displayed for years on the lawn of the Science and Engineering Library at SMU. In repairing this more than 1,000-pound track, Bill Lowe and I had to rotate it several times. We gained a lot of respect for Bird and his 1940s workers in Glen Rose. Only those who have dealt with these large tracks in thick, heavy rock can appreciate how difficult even one is to handle. Imagine trying to gift wrap an automobile. That is the general idea.

A carnivore track from Glen Rose is on exhibit at the Strecker Museum at Baylor University and another at the Dallas Museum.

Texas Has Many Dinosaur Tracks

A very important point to be made about dinosaur tracks is that Glen Rose and the Paluxy River have no monopoly on Texas dinosaur tracks. The Glen Rose limestone is widely exposed by erosion in central Texas. Dozens of Texas ranches have these 115-million-year-old footprints. At that period in prehistory, central Texas was experiencing one of several Cretaceous inundations by the waters of the Gulf of Mexico. The shoreline stood about where the Paluxy River flows today in Somervell County near Glen Rose—that is considerably inland from the present Texas coast. A broad, shallow coastal lagoon allowed dinosaurs to cavort and feed in the shallow water, leaving footprints in the limey muds, a perfect physical setting for making many fossil trackways. The second ingredient needed to preserve the tracks is a gentle filling-in of the prints by sediment to keep their shape intact. It was sort of a rock sandwich. Millions of years of burial and eventual erosion back to the surface are needed to complete this cycle. We see the results of this today in the bed of the Paluxy River and many other places in central Texas.

A Spectacle in Arkansas

Dr. Jeffrey Pittman, a fossil trackway expert, took a wonderful photograph in Arkansas of rock similar to that found in Texas. The view from a low-flying airplane shows hundreds of sauropod dinosaur tracks completely covering an area bigger than a football field. The tracks are similar to those *Brontopodus* tracks, usually blamed on the dinosaur *Pleurocoelus*. This snapshot must record the passage of a large herd of huge (up to 60-foot-long) dinosaurs. A close examination reveals this to be a truly fantastic photograph.

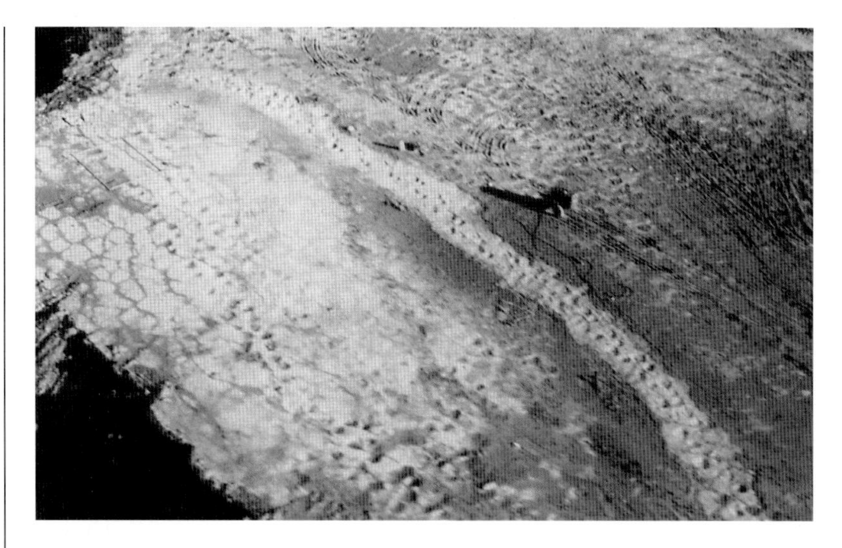

Dr. Pittman's own caption for this remarkable picture reads: "Hundreds of sauropod dinosaur tracks were found on a limestone layer in an Arkansas gypsum mine. This photograph of a football-field-sized area was taken from an airplane. An air compressor (half the size of a car) is in the center of the picture. Near the air compressor is a long sauropod trackway cleaned of mud and rock. Several other trackways are visible near the quarried edge of the exposure; many more are visible beyond the compressor." These tracks were probably made by a herd of Pleurocoelus dinosaurs. Photo by Dr. Jeffrey Pittman.

The Fort Worth Museum of Science and History

Until the Fort Worth Museum completes its current efforts with SMU to reconstruct the skeletons of the big brachiosaurid *Pleurocoelus* or three-toed meat-eater *Acrocanthosaurus,* the skeletons of these wonderful Texas dinosaurs that most probably made the famous Glen Rose trackways are nowhere on exhibit. It should be noted that a cast of an Oklahoma *Acrocanthosaurus* from McCurtain County, Oklahoma, has just been completed by a northern exhibit firm. Perhaps it will appear in the new Oklahoma Museum of Natural History in Norman, Oklahoma. After all, that is where Drs. Stovall and Langston first described that toothy creature.

Better Indicators of Ancient Behavior

Trace fossils have a big advantage over mere bones when it comes to revealing ancient behavior. Early on, R. T. Bird imparted behavioral interpretations to trackways at Glen Rose. His most famous example came when he found the parallel paths of a large sauropod, probably *Pleurocoelus,*

Trackway at Glen Rose excavated in 1940 by R. T. Bird. Half of it is on display at the American Museum of Natural History, New York. The bathtub-like prints are (track name) Brontopodus birdi, *ascribed to the dinosaur* Pleurocoelus. *The three-toed tracks are (track name)* Irenesauripus glenrosensis, *ascribed to the dinosaur* Acrocanthosaurus. Photo 324393, courtesy of the Department of Library Services, American Museum of Natural History.

and the smaller three-toed trail of probably a carnivorous *Acrocanthosaurus*. Bird was naturally forced to speculate on real events that might have occurred 110 million years earlier in the everyday life of the two dinosaurs. He quite logically concluded that the meat-eater was up to no good and, in all likelihood, was attacking the big herbivore. That is how tracks can help us "get to know" the dinosaurs in ways that bones cannot.

Zilker Park in Austin

A rock quarrying effort in Zilker Park, just across the Colorado River from downtown Austin, Texas, uncovered a two-acre area of six-to-twelve-inch dinosaur tracks. Carbon preservations of once waving water reeds were also present. A fossil turtle was found, which probably was at home with the dinosaurs in this long-gone lagoon about 110 million years ago. The site showed long trails of tracks of small-to-medium-sized dinosaurs, probably fishing among the reeds. The Zilker tracksite has been examined by one of the leading authorities on Texas dinosaur tracks, Dr. Jeff Pittman of the University of Colorado at Denver. He

A plaster cast of one of the grallatorid tracks found in Zilker Park in Austin, Texas, from the Vertebrate Paleontology Lab, UT Austin.

has tentatively identified the tracks as "grallatorid," apparently similar in footprint shape to the flesh-eating dinosaurs that left some tracks described as from a very fast-moving dinosaur by Dr. James Farlow. Dr. Farlow's work near Junction, Texas, will be discussed later. In the case of the Zilker Park trackways, Dr. Pittman has estimated these grallatorids were lazing-along at only 4 to 7 mph. A leisurely fishing trip in the reed beds! Remember, tracks can look similar even when there is a good chance that probably two different dinosaur species are the makers.

Grapevine Lake Trackways

In 1982, a geology field trip from Brookhaven College in Dallas visited the spillway of Grapevine Lake just after a huge rainstorm. Many layers of the Upper Cretaceous Woodbine Formation had been stripped away, displaying at least two different dinosaur trackways. The tracks were in rock about 95 million years old, almost 20 million years younger than the Glen Rose tracks. One of these trackways was removed by Bob Slaughter of SMU's Shuler Museum and the other by crews led by William (Bill) Wilson and Steve Runnels of the Dallas Museum of Natural History.

Professor Bob Slaughter (far left) with Dr. David Gillette (lower left) and several students. This picture was taken during work on dinosaur tracks at Lake Grapevine.

Ghost Prints?

The Grapevine Lake site was my first introduction to "ghost" prints. It was near Halloween, come to think. One of the Lake Grapevine spillway trackways was eight inches deep, very clear, with sharp edge definition; the other (about 100 yards away and not the same trail) was composed of shallow, less distinct tracks. The sediment consisted of a muddy sandstone laid down in layers.

Dr. Langston and others have written about the way, under certain circumstances, a heavy dinosaur will leave a print not only on the layer upon which it actually steps, but also in lesser and lesser amounts on each separate layer underneath. Apparently this can penetrate three or four layers, leaving less and less of a print on each layer. Besides the "real" tracks you get several layers with "ghost" prints. In addition, an original print can be partially filled with a later sediment layer or two and still show a fainter and somewhat altered track in those layers above. This could cause one dinosaur to be identified as several very different species. When only an isolated track is found on a site, it is hard to know if it is an original or ghost print. Tracks are indirect evidence, and it is difficult to interpret them definitively.

Remember, a footprint really is a fossil. It can be given a name even though its maker is unknown or uncertain. Somewhere else a dinosaur bone is given a different name, although it may be the same kind of dinosaur that made the track. Putting tracks and bones together is a difficult and uncertain business. Fortunately, Gary Byrd's hadrosaur, *Protohadro byrdi,* was found not far from these Lake Grapevine tracks. Until that discovery, the Lake Grapevine tracks had been variously thought to be either hadrosaur or iguanodont. Now we can guess that some of the tracks are hadrosaurian, the duckbill family.

Pigeon-Toed Dinosaurs?

The second Lake Grapevine dinosaur track site is under the lake and seldom, if ever, viewable. The tracks represent at least three kinds of dinosaurs, with hadrosaurs (duckbills) being most plentiful. In places, the tracks are so plentiful that they cover up each other, where several creatures seem to have milled around. My friend and helper William (Bill) Lowe lightly researched the extremely "pigeon-toed" habit shown by the hadrosaur footprints on this site. Although pigeon-toed trails are known, Bill found no mention of a trail as severely pigeon-toed anywhere in the literature. The tracks remind me very much of the way a heron paces in shallow water, angling in the

Probable hadrosaur (duck-billed dinosaur) tracks found near Lake Grapevine, Woodbine Formation. A trail of similar-sized tracks is going in both directions. Some dinosaur bones are occasionally found in the area.

toes of both feet to scare its prey out of hiding in the bottom sediment. Many dinosaur trails pace off rather evenly, but these are quite varied in stride, almost a dance step. Can you imagine highly educated paleontologists, miles from anywhere on an isolated fossil track site, bending and staggering, trying to imitate a dance step of the dinosaurs?

The sediment in this part of the Woodbine Formation shows a sandy lagoonal environment with indications of considerable worn seashell and shale accumulation revealing a frequently changing depth. There is much fossil wood in the Woodbine Formation, some of it hollowed by tubes of wood-boring bivalves. These rocks were deposited on the eastern edge of the encroaching sea that had formerly, and would soon again, cover all of Texas. This northeastern corner of Texas warrants close examination. Within only the last few years, we have seen evidence that it may have come out of the Cretaceous sea more often than has been supposed. There is some evidence that a volcanic island, similar to that found at Pilot Knob near Austin, Texas, may have existed in north Texas as well. Some paleogeographic maps show such volcanic features not far north of the Dallas-Fort Worth area.

Faster Than a Speeding Tyrannosaur?

Dr. James Farlow, of Indiana University-Purdue at Fort Wayne, has spent a lot of time in Texas studying dinosaur footprints, trying to determine how fast various dinosaurs could walk and run. On a ranch near Junction, Texas, Dr. Farlow has clocked his "world's fastest dinosaur" at over 26 mph. It was a small carnivorous dinosaur known mostly from its footprints, which have been given the footprint name of *Grallator.* Human athletes can briefly attain a speed of about 22 mph. So don't even think of outrunning a dinosaur, at least, not that one!

Dr. James Farlow has studied the modern walking and running habits of birds, which are very closely related to dinosaurs, to determine what behaviors he can deduce from modern and fossil trackways. Large birds that rely on flightless locomotion are naturally his favorites. The length of legs and distance apart of the tracks are the keys to such determinations.

A young friend of mine, Annette Buckley, used Dr. Farlow and Langston's generous help and their studies of dinosaur trackways to do a school project to determine just how fast a *Tyrannosaurus* can run. We have seen these ferocious meat-eaters chasing down Jeeps in recent movies. Annette concluded, with the help of the experts, that T-rex probably had a top speed of about 22 mph. So the best human runners could almost outrun a tyrannosaurus. The good news for T-rex was that most of his prey was slower than we are!

That's our Story

We have traced many of the highlights of dinosaur study in Texas. More and more research is being completed and more and more field discoveries are being made each year. Dinosaur work in Texas has been the work of relatively few people. This book has introduced you to many of them. Several Texas museums are currently working on dinosaur fossils for research and for exhibit. One thing that never seems to change is people's fascination with dinosaurs. I sincerely hope this book will pique your fascination and be an aid to your study.

Bibliography

Adkins, Walter S., "The Mesozoic Systems of Texas," in *The Geology of Texas,* v. 1, University of Texas Bulletin 3232, 1933.

Allen, Tom and Jane D. Allen, *Dinosaur Days in Texas,* Dallas: Hendrick-Long Publishing Co., 1989.

Alvarez, L. W., Walter Alvarez, Frank Asaro, and H. V. Michel. "Extraterrestrial Cause for the Cretaceous-Tertiary Extinction," *Science,* v. 208, 1980, pp. 1095–1108.

Andrews, Roy Chapman, *Under a Lucky Star, A Lifetime of Adventure,* New York: Viking Press, 1943.

Bird, Roland T. (V. Theodore Schreiber, ed.), *Bones for Barnum Brown, Adventures of a Dinosaur Hunter,* Fort Worth: Texas Christian University Press, 1985.

Bourgeois, J., Thor A. Hansen, P. L. Wiberg, and E. G. Kauffman, "A Tsunami Deposit at the Cretaceous-Tertiary Boundary in Texas," *Science,* v. 241, 1988, pp. 561–570.

Busbey, Arthur B. and Thomas M. Lehman, eds., *Vertebrate Paleontology, Biostratigraphy, and Depositional Environments, Latest Cretaceous and Tertiary, Big Bend Area, Texas* (Guidebook), 49th Annual Meeting, Society of Vertebrate Paleontology, Austin, Texas, 1989.

Chatterjee, Sankar, "The Late Triassic Dockum Vertebrates," in *The Beginning of the Age of Dinosaurs,* Kevin Padian, ed., Cambridge, U.K.: Cambridge University Press, 1986, pp. 139–150.

Chatterjee, Sankar, "The Dawn of the Age of Dinosaurs," in *The Ultimate Dinosaur,* Byron Preiss et al., eds., New York: Bantam Press, 1992, pp. 2–9.

Chatterjee, Sankar, "Shuvosaurus, A New Theropod," in *National Geographic Research and Exploration,* v. 9 (3), 1993, pp. 274–285.

Currie, Phillip J. and Kevin Padian, eds., *Encyclopedia of Dinosaurs,* New York: Academic Press, 1997.

Davies, Kyle Linton, *Hadrosaurian Dinosaurs of the Big Bend National Park, Brewster County, Texas,* unpublished master's thesis, University of Texas at Austin, August, 1983.

Dingus, Lowell and Timothy Rowe, *The Mistaken Extinction, Dinosaur Evolution and the Origin of Birds,* New York: Freeman and Company, 1998.

Dott, Robert H., Jr. and Donald R. Prothero, *Evolution of the Earth, Fifth Edition,* New York: McGraw-Hill, 1994.

Farlow, James O., "Estimates of Dinosaur Speeds from a New Trackway Site in Texas," *Nature,* v. 294, 1981, pp. 747–748.

Farlow, James O. and M. K. Brett-Surman, eds., *The Complete Dinosaur,* Indiana Press, 1997.

Forster, Catherine A., Paul C. Sereno, Thomas W. Evans, and Timothy Rowe, "A Complete Skull of *Chasmosaurus mariscalensis* from the Aguja Formation (Late Campanian) of West Texas," *Journal of Vertebrate Paleontology,* v. 13 (2), 1993, pp. 161–170.

Gillette, David G. and Martin G. Lockley, eds., *Dinosaur Tracks and Traces,* Cambridge, U.K.: Cambridge University Press, 1989.

Hildebrand, Alan R., *et al.,* "Chicxulub Crater: A Possible Cretaceous/Tertiary Boundary Impact Crater on the Yucatan Peninsula, Mexico," *Geology,* v. 19, pp. 867–871.

Hill, Robert T., "Geography and Geology of the Black and Grand Prairies, Texas," *U.S. Geological Survey, 21st Annual Report,* Washington, D.C., 1901, p. 7.

Jacobs, Bonnie Fine, "Paleobotany of the Lower Cretaceous Trinity Group, Texas," in *Field Guide to the Vertebrate Paleontology of the Trinity Group, Lower Cretaceous of Central Texas,* Winkler, Dale A. et al. eds., 49th Annual Meeting, Society of Vertebrate Paleontology, SMU, 1989, pp. 31–33.

Jacobs, Louis L. , Dale A. Winkler, Phillip A. Murry, and John M. Maurice, "A Nodosaurid Scuteling from the Texas Shore of the Western Interior Seaway," in *Dinosaur Eggs and Babies,* Kenneth Carpenter and John R. Horner, eds., Cambridge, U.K.: Cambridge University Press, 1994, pp. 337–346.

Jacobs, Louis, *Lone Star Dinosaurs,* College Station: Texas A&M Press, 1995.

Langston, Wann, Jr., "Non-mammalian Comanchean Tetrapods," in *Geoscience and Man,* v. 8, Baton Rouge: Louisiana State University, 1974, pp. 77–102.

Lawson, Douglas A., "*Tyrannosaurus* and *Torosaurus,* Maestrichtian Dinosaurs from Trans-Pecos, Texas," *Journal of Paleontology,* v. 50(1), 1976, pp. 158–164.

Lee, Yuong-Nam, "A New Nodosaurid Ankylosaur (Dinosauria: Ornithischia) from the PawPaw Formation (Late Albian) of Texas, *Journal of Vertebrate Paleontology,* v. 16 (2), 1996, pp. 232–245.

Lehman, Thomas M., "Late Maastrictian Paleoenvironments and Dinosaur Biogeography in the Western Interior of North America," *Palaeogeography, Palaeoclimatology, Palaeoecology,* v. 60, 1987, pp. 189–217.

Lehman, Thomas M., "*Chasmosaurus mariscalensis,* sp. nov., A New Ceratopsian Dinosaur from Texas," *Journal of Vertebrate Paleontology,* v. 9 (2), 1989, pp. 137–162.

Lehman, Thomas M., "Late Campanian Dinosaur Biogeography in the Western Interior of North America," *Dinofest,* 1997.

McGowen, Mary K. and Cynthia M. Lopez, *Depositional Systems in the Nacatosh Formation (Upper Cretaceous), Northeast Texas and Southwest Arkansas,* Report of Investigations No. 137, Bureau of Economic Geology, UT Austin, 1983.

Shuler, Ellis W., "Dinosaur Tracks at the Fourth Crossing of the Paluxy River near Glen Rose, Texas," in *Field and Laboratory,* v. 5, SMU, Dallas, 1937, pp. 33–36.

Stovall, J. Willis and Wann Langston, Jr., "*Acrocanthosaurus atokensis,* a New Genus and Species of Lower Cretaceous Theropoda from Oklahoma," *American Midland Naturalist,* v. 43, 1950, pp. 696–728.

Tickner, Bruce, *Stratigraphic Studies and Microfacies Analysis, of the Jurassic Succession, Malone Mountains, West Texas,* master's thesis, University of Texas at Dallas, December 1987.

Vance, Thomas, *Prehistoric Reptiles of Texas (Annotated Checklist and Bibliography, 1860–1995),* Corsicana, Texas: Navarro College, unpublished.

Wheeler, E. A., Thomas M. Lehman, and P. E. Gasson, "*Javelinoxylon,* An Upper Cretaceous Dicotyledonous Tree from Big Bend National Park, Texas, with Presumed Malvalean Affinities," *American Journal of Botany,* v. 81 (16), 1991, pp. 703–710.

Wilcox, R. E. and Wann Langston, Jr., and S. C. Good, "The Nonmarine Bissett Formation of the Western Glass Mountains, Trans-Pecos Texas: Evidence for an Early Cretaceous (Neocomian(?)–Albian) Age," in *South-Central and Southeast Section, Geological Society of America Abstract with Programs,* 1986, p. 272.

Wilcox, Robert E., *Evidence for an Early Cretaceous Age for the Bissett Formation, Western Glass Mountains, Brewster and Pecos Counties, West Texas: Implications for Regional Stratigraphy, Paleogeography, and Tectonics,* master's thesis, Alpine, Texas: Sul Ross State University, April 1989.

Wilcox, Robert and Dr. David M. Rohr, "Evidence for a Cretaceous Age for the Bissett Formation," in *West Texas Geological Society Bulletin,* v. 35, no. 7, March 1996, pp. 5–12.

Winkler, Dale A. and Phillip A. Murry, "Paleoecology and Hypsilophodontid Behavior at the Proctor Lake Dinosaur Locality (Early Cretaceous), Texas," *Geological Society of America, Special Paper,* v. 238, 1989, pp. 55–61.

Winkler, Dale A., Phillip A. Murry, and Louis L. Jacobs, "Early Cretaceous (Comanchean) Vertebrates of Central Texas," *Journal of Vertebrate Paleontology,* v. 10, 1990, pp. 95–116.

Winkler, Dale A., Phillip A. Murry, Louis L. Jacobs, J. Russell Branch and Patrick Trudell, "The Proctor Lake Dinosaur Locality, Lower Cretaceous of Texas," *Hunteria,* v. 2 (5), 1988, pp. 1–8.

Yancy, Thomas E., ed. and Karl J. Koenig, *The Cretaceous-Tertiary Boundary and Lower Tertiary of the Brazos River Valley,* San Antonio: South Texas Geological Society, 1984.

Index